Developing Embedded Software using DaVinci & OMAP Technology

Synthesis Lectures on Digital Circuits and Systems

Editor
Mitchell A. Thornton, ***Southern Methodist University***

Developing Embedded Software using DaVinci & OMAP Technology
B.I. (Raj) Pawate
2009

Mismatch and Noise in Modern IC Processes
Andrew Marshall
2009

Asynchronous Sequential Machine Design and Analysis: A Comprehensive Development of the Design and Analysis of Clock-Independent State Machines and Systems
Richard F. Tinder
2009

An Introduction to Logic Circuit Testing
Parag K. Lala
2008

Pragmatic Power
William J. Eccles
2008

Multiple Valued Logic: Concepts and Representations
D. Michael Miller, Mitchell A. Thornton
2007

Finite State Machine Datapath Design, Optimization, and Implementation
Justin Davis, Robert Reese
2007

Atmel AVR Microcontroller Primer: Programming and Interfacing
Steven F. Barrett, Daniel J. Pack
2007

Pragmatic Logic
William J. Eccles
2007

PSpice for Filters and Transmission Lines
Paul Tobin
2007

PSpice for Digital Signal Processing
Paul Tobin
2007

PSpice for Analog Communications Engineering
Paul Tobin
2007

PSpice for Digital Communications Engineering
Paul Tobin
2007

PSpice for Circuit Theory and Electronic Devices
Paul Tobin
2007

Pragmatic Circuits: DC and Time Domain
William J. Eccles
2006

Pragmatic Circuits: Frequency Domain
William J. Eccles
2006

Pragmatic Circuits: Signals and Filters
William J. Eccles
2006

High-Speed Digital System Design
Justin Davis
2006

Introduction to Logic Synthesis using Verilog HDL
Robert B.Reese, Mitchell A.Thornton
2006

Microcontrollers Fundamentals for Engineers and Scientists
Steven F. Barrett, Daniel J. Pack
2006

Developing Embedded Software using DaVinci & OMAP Technology

B.I. (Raj) Pawate

www.morganclaypool.com

ISBN: 9781598299786 paperback
ISBN: 9781598299793 ebook

DOI 10.2200/S00189ED1V01Y200904DCS021

A Publication in the Morgan & Claypool Publishers series
SYNTHESIS LECTURES ON DIGITAL CIRCUITS AND SYSTEMS

Lecture #21
Series Editor: Mitchell A. Thornton, *Southern Methodist University*

Series ISSN
Synthesis Lectures on Digital Circuits and Systems
Print 1932-3166 Electronic 1932-3174

DISCLAIMER
Neither the author nor the publisher guarantees the accuracy and completeness of the information published herein. Neither the publisher nor the author shall be responsible for any errors, omissions, or damages arising out of use of this information. The sample code provided in this book is intended only as a general guide and there might be other and better ways of developing software. No information provided in this book is intended to be or shall be construed to be an endorsement, certification, approval, recommendation, or rejection of any particular supplier, product, application, or service.

Developing Embedded Software using DaVinci & OMAP Technology

B.I. (Raj) Pawate
Texas Intruments Incorporated

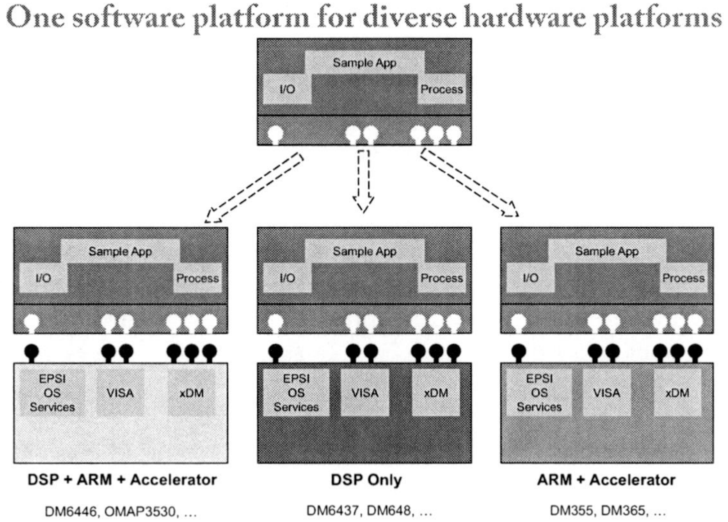

SYNTHESIS LECTURES ON DIGITAL CIRCUITS AND SYSTEMS #21

MORGAN & CLAYPOOL PUBLISHERS

ABSTRACT

This book discusses how to develop embedded products using DaVinci & OMAP Technology from Texas Instruments Incorporated. It presents a single software platform for diverse hardware platforms. DaVinci & OMAP Technology refers to the family of processors, development tools, software products, and support. While DaVinci Technology is driven by the needs of consumer video products such as IP network cameras, networked projectors, digital signage and portable media players, OMAP Technology is driven by the needs of wireless products such as smart phones.

Texas Instruments offers a wide variety of processing devices to meet our users' price and performance needs. These vary from single digital signal processing devices to complex, system-on-chip (SoC) devices with multiple processors and peripherals. As a software developer you question: Do I need to become an expert in signal processing and learn the details of these complex devices before I can use them in my application? As a senior executive you wonder: How can I reduce my engineering development cost? How can I move from one processor to another from Texas Instruments without incurring a significant development cost? This book addresses these questions with sample code and gives an insight into the software architecture and associated component software products that make up this software platform. As an example, we show how we develop an IP network camera.

Using this software platform, you can choose to focus on the application and quickly create a product without having to learn the details of the underlying hardware or signal processing algorithms. Alternatively, you can choose to differentiate at both the application as well as the signal processing layer by developing and adding your algorithms using the xDAIS for Digital Media, xDM, guidelines for component software. Finally, you may use one code base across different hardware platforms.

KEYWORDS

signal processing, system-on-chip (SoC), eXpressDSP Algorithm interface standard, xDAIS, xDAIS for Digital Media, xDM, component software, code base, hardware platform, software platform, DaVinci Technology, OMAP Technology, video codecs, sample application.

To my wife Parvathi and our children Varsha, Veena, Nidhi;

And

to the people at Texa Intruments Incorporated
who created and launched the TMS32010,
a digital signal processor.

Contents

Foreword

When TI began our drive into the world of Digital Signal Processing, life was easy. We had very simple processing engines (TMS32010, TMS32020). Those processors didn't have a lot of memory, nor did they have many peripherals or standard I/Os. Certainly, they didn't have things like pipelines, interrupts or cache. Their instruction rates were fast for the time, but still slow compared to what we see now (five to ten million instructions per second).

As an historical aside, our first architecture, the TMS32010, initially had no interrupts. It was quite correctly assumed that real time signals are not interruptable and therefore a processor designed to process real time signals should not be interrupted. After a great amount of lively discussion, it was concluded that an interrupt mechanism needed to be added to the architecture. But, it was designed such that the processor could effectively interrupt the interrupt until it was through with its real time task. Now, back to the discussion.

The problems we were trying to solve then didn't take a lot of programming as they were basically filters or transforms which could be executed in no more than a hundred or so assembly level instructions. So, they were fairly simple to do.

But then as the design community began to adopt this new microcontroller concept, they added two conflicting demands: (1) More performance, and (2) Easier to use. With these demands we methodologically made the DSP more complex and, at the same time, hid these complexities with a development environment. At the same time, the number of lines of code grew orders of magnitude greater than the thousands of lines of code.

So, in today's world of Signal Processing, we find ourselves writing our code in Matlab, C or some sort of meta-language. This gives us the ease of use (mostly), but has forced us to significantly over deliver on performance – yes, to make up for the inefficiencies involved with ease of use. But, with the inefficiency we have lost the opportunity to take full advantage of the raw performance of the processor, or processors.

That is where a book of this sort comes in handy. There are many ways to take more advantage of the raw performance without having to learn to write it all in assembly language. The pages that follow will help you find manageable ways to get to market fast while still grabbing more of the processors raw performance. Specifically this book addresses three very important aspects of writing code: (1) a base software platform with optimized signal processing algorithms/codecs, (2) system level performance assessment and tuning, and (3) scalability. All of this on a software platform on which developers can build.

Gene A. Frantz, *Principal Fellow*
Texas Intruments Incorporated

Preface

During the latter part of 2004, Texas Instruments decided to offer rigorously tested software components such as video codecs as software products. This was a significant decision since the company previously was accustomed to selling silicon devices, and evaluation modules, with development tools and some demonstration software. The demonstration software was included as a starter ware with the intent of showing some of the capabilities of the processor. However, real time signal processing solutions have long transitioned from being a few hundred lines of code to several hundred thousand lines of code. The devices themselves have become quite complex, with multiple levels of cache, peripherals, and processors on one silicon die. A need exists to provide a robust software platform that abstracts the complexities of the device and signal processing algorithms on top of which users can build their applications. At the same time, users should be able to go deep into the device if they so desire.

I have had the pleasure of working with several people at Texas Instruments in putting together the strategy and process for delivering these software products and creating a software architecture that would show how to use these components and build an application. No strategy is good unless it is put into practice and lessons learned from it. After evangelizing this software architecture for a while, starting 2006, I focused on an emerging market in video surveillance, in particular on digital video recorders and IP network cameras. During this time, we interacted with several customers who leveraged this software platform, created products and ramped them to production.

During this phase, I realized that while we had a wealth of documentation on specific topics regarding DaVinci Technology, we did not have one big picture document or book that addressed the software platform as a whole with some examples. As we interacted with our users, based on some of the typical questions, I would create power point presentation slides and then several of us would help our users solve their problems and get to market.

One of the important things that we did was to work towards a single software platform so that developers would have the same look and feel and development experience whether they used a system-on-chip with an ARM and DSP and accelerator or just an ARM with an accelerator.

This book has been in the making since the latter part of 2006. I hope it meets the need of a single document that discusses what the software component products are, how they are related to each other, how users may build on top of them, and move from one type of hardware platform to another.

B.I. (Raj) Pawate
Houston, Texas
March 27, 2009

Acknowledgement

This book has been "cooking" for quite some time, given that the focus was more on getting our users to production using our devices, that it took me several years. I wish to thank Cathy Wicks, University Program Manager, Texas Instruments, for talking me into writing this book. Also thanks to Phillip Parker, Market Communications Manager, Texas Instruments, for providing me with assistance to get this book done.

I came to know Gene Frantz when I joined TI in the speech group and he has been my mentor for over 20 years. I'd like to thank him for encouraging me and providing guidance all these years and also for writing the foreword.

A special thanks to Girish Kanmas who helped me with this document, formatting, drawing the figures and putting together the first version of this book.

I also would like to acknowledge Aravindhan Karupiah and Anand Balagopalakrishnan, Texas Instruments India Pvt. Ltd., for contributing to two chapters in this book. I have referred to them in the appropriate chapters. That lunch discussion in Indira Nagar, Bangalore was well worth it! A special thanks to Ajit Rao and the Multimedia Codecs development team in Texas Instruments India, Bangalore for taking xDM to the next level. The Digital Video Test Bench (DVTB) code was developed for internal test and debug, but during the course of getting some of our early adopters to production, we started releasing it and eventually it became a part of our Digital Video Software development kit (DVSDK). Thanks to Zahid Qamar and Anand Balagopalakrishnan for making this happen. While several products have been built using this software platform, we discuss the development of an IP Network camera as an example application built on top of the DVSDK. Thanks to Cheng Peng, Raghavendra Kudva, Fitzgerald Archibald, Wen Chi, Chris Ring, and Senthil Natarajan for contributing to this chapter.

I'd like to give a special thanks to the TI Technical Organization, in particular Scott Specker, Tim Cartier, and Steve Preissig for evangelizing this software architecture and platform. They have put together excellent training materials and conducted sessions throughout the world.

I also wish to thank Steve Ling and Tim Adcock, Field Applications Managers, both with Texas Instruments Inc. for encouraging me to write an executive summary as a good tradeoff between the book for software developers who wish to see some reference code and senior executives who wish to quickly get a summary of the book.

My career in signal processing started with the TMS32010 and hence I felt I should dedicate this book to the people at Texas Instruments that created and marketed this device successfully.

Special thanks to Danny Petkevich and Jean-Marc Darchy for reviewing and encouraging me to complete this work.

My family, including my sister Girija Mahant, and my brothers Prof. S.I. Pawate, Dr. C.I. Pawate, Veer Pawate, and Prabhu Pawate have always been supportive and encouraging since my childhood days and I would like to express my thanks to them.

Finally, I wish to thank Joel Claypool and Dr. C.L. Tondo at Morgan–Claypool for their enthusiasm which guided this project to completion.

Abbreviations

Abbreviation	Description
3A	AutoFocus, AutoWhiteBalance, AutoExposure
3P	TI Third Party Network Company
AC1	Hardware Accelerator 1
AC2	Hardware Accelerator 2
API	Application Programming Interface
APL	Application Layer
ARM	Advanced RISC Machines Ltd
ARM9EJ	ARM 32-bit RISC Processor
CE	Codec Engine
DMA	Direct Memory Access
DMAN3	DMA Manager
DSP	Digital Signal Processing
DSPLIB	General signal processing library for C64x+ DSP
DVSDK	Digital Video Software Development Kit
DVTB	Digital Video Test Bench
EDM	EPSI to Driver Mapping
EPSI	Embedded Peripheral Software Interface
EVM	Evaluation Module
HD VICP	High Definition VICP
IMGLIB	Image processing library for C64x+ DSP
iMX	Hardware accelerator on DM355 device
IOL	Input Output Layer
IP	Internet Protocol
IPNC	IP Network Camera
SoC	System-on-chip
SPL	Signal Processing Layer
VALIB	Video Analytics library running on C64x+ DSP
VICP	Video, Imaging Coprocessor
VISA	Video, Imaging, Speech & Audio
VPBE	Video Port Back End
VPFE	Video Port Front End
VPSS	Video Port Sub System
xDAIS	eXpressDSP Algorithm Interface Standard
xDM	eXpressDSP Digital Media

Executive Summary

Product developers like to preserve their investment in a platform and create many spin-offs or derivative products from it as long as possible. But new features, product evolution, and competition require transition to next generation platforms. Texas Instruments strives to provide technological innovations and rapidly introduces best in class solutions at different price-performance points. There are typical questions that may run through your mind as a business or engineering manager. How can you leverage these offerings and rapidly introduce new products? Can you replace functions in your system with best in class components developed either by yourself or by third parties (3Ps) without affecting the rest of the system? Do you need to become an expert in signal processing before you use these devices?

In order to address these questions, TI introduced a software platform and development kit in 2005. Since then several customers have adopted this platform, developed products quickly and ramped them to production. Examples include, but not limited to, video security IP cameras, digital video recorders, networked video recorders, networked projectors, media players, and so on. The rest of the document discusses this methodology and explains with example code how a software developer might leverage this rich set of hardware-software offerings from Texas Instruments into a product.

Texas Instruments offers a wide variety of processing devices to meet users' different price and performance needs. These devices can be broadly classified into three types:

1. DSP + ARM + Accelerator system-on-chip (SoC)

2. DSP only

3. ARM + Accelerator SoC

The first type integrates the well known C64x+ DSP core with a general purpose processor (GPP) ARM9 coupled with dedicated hardware accelerators. It offers users the flexibility of a GPP coupled with a programmable DSP as well as dedicated accelerators. The DM6446 and the OMAP353x are examples of these type of devices. For cost sensitive, low power applications, the second type integrates a C64x+ core DSP with a rich set of peripherals. An example device is the DM6437. For dedicated, cost sensitive applications, the third type integrates an GPP with an accelerator. Examples include the DM355, a MPEG4 encode and decode device and the DM365, a H.264/MPEG4 encode and decode device.

ONE SOFTWARE PLATFORM FOR DIVERSE PRICE-PERFORMANCE HARDWARE PLATFORMS

There are several benefits to using the scalable TI software framework shown in Figure 1:

- ☐ Reduce time to market
- ☐ Focus on differentiation while leveraging the rest of the system from TI
- ☐ Easily replace existing IP with best-in-class components
- ☐ Easily migrate to a different price-performance hardware platform from TI

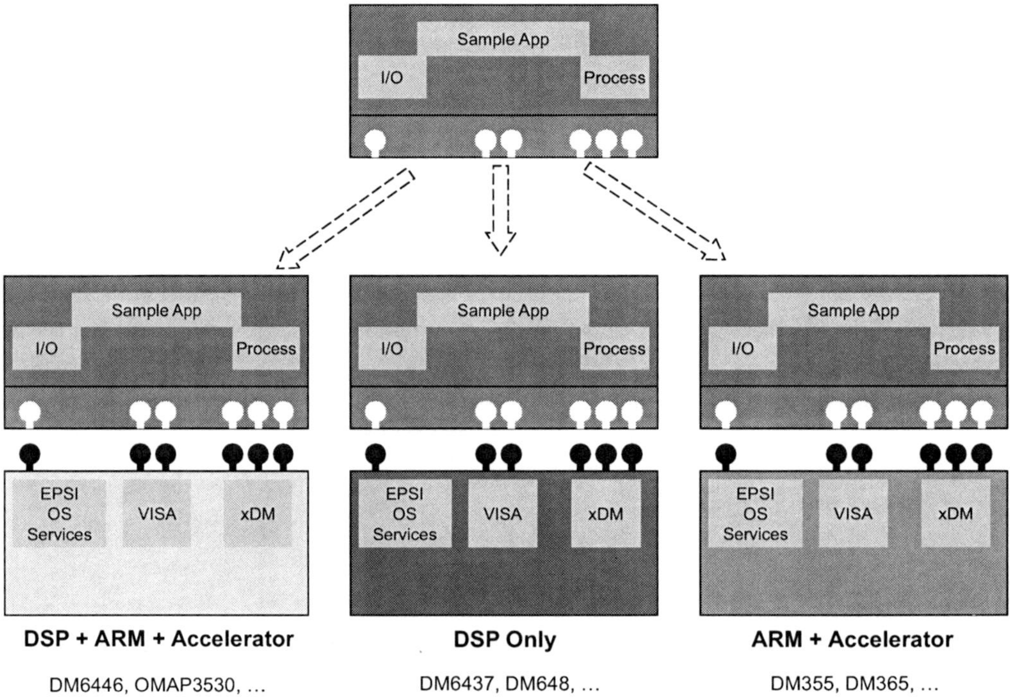

Figure 1: One software platform for diverse price-performance hardware platforms.

DAVINCI & OMAP TECHNOLOGY

DaVinci Technology refers to the "DM" platforplatform of media processors with their associated development tools, software components, and support infrastructure including third party companies. These devices were driven by video market requirements such as set-top box, video phone, digital still cameras, etc. Some examples are the DM6446, DM6437, DM6467, DM355, etc.

OMAP Technology refers to the "OMAPxxxx" platform of processors with their associated development tools, software components, and support infrastructure including third party companies. These devices were driven by wireless market requirements. Some examples are the OMAP353x, OMAPL1x, etc.

TI SOFTWARE ARCHITECTURE AND STACK

Figure 2 shows the TI software architecture designed to support the use case of reading or inputting a buffer of data, processing it, and outputting a buffer of data. A "sample application" shows the developer how to call the two main APIs, input-output APIs and process APIs. Software functions or programs are grouped into three main layers, the Application Layer (APL), the Input-Output layer (IOL), and the Signal Processing Layer (SPL). Depending on the type of silicon device, all of these three layers may run on one device, for example the DM6437; in other devices such as OMAP353x or DM6446, the Application layer and Input-Output layer are partitioned to run on the ARM while the Signal Processing layer runs on the C64x+ DSP.

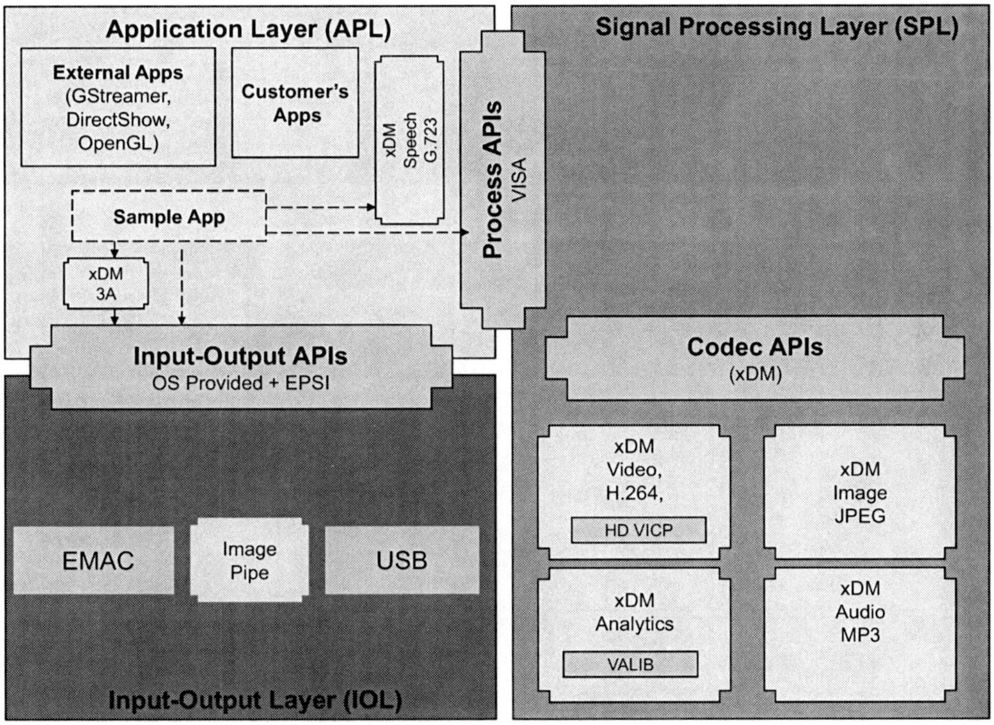

Figure 2: TI Software Architecture.

As the names imply, the input-output layer is primarily associated with functions performing input and output. It consists of drivers integrated into an Operating System such as Linux, or WinCE. The Signal Processing Layer is primarily associated with processing data and consists of all the compute intensive, usually signal processing functions such as H.264, JPEG, etc. The application layer consists of all high level user interface functions, networking stack, and depending on the capability of the silicon platform, may also implement some of the signal processing functions.

xDM, which stands for xDAIS for Digital Media, is a standard wrapper that is put around all processing functions such as H.264 encode or G.729 codec. Adhering to the development guidelines as outlined in xDAIS and xDM and making your code compliant with the coding guidelines from TI, and finally following this software architecture, enables software to be ported from one platform to another as well as allowing easy replacement of the components.

Figure 3 shows the software stack. At the foundation is the hardware consisting of the peripherals, processing elements such as ARM, DSP, and one or more hardware accelerators AC1, AC2, etc. Running on the ARM and utilizing the peripherals is the Operating System.

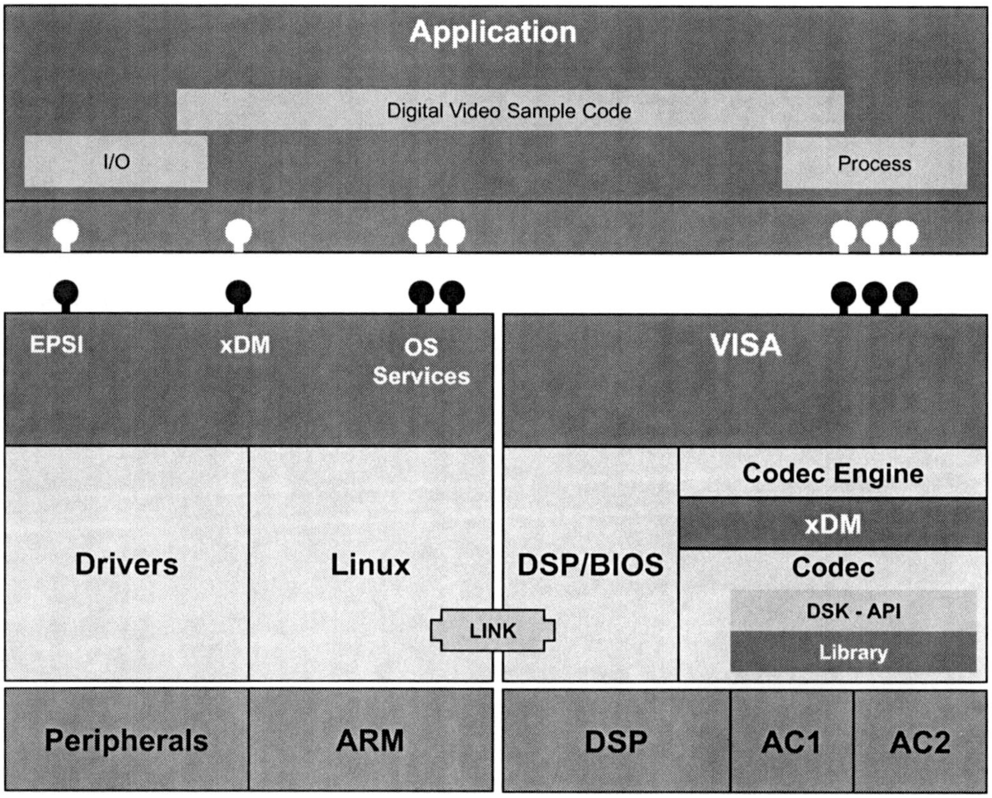

Figure 3: Software Stack.

Customers can develop software in the application layer and leverage third party company (3P) and opensource software where applicable. Any higher level APIs such as OpenMax, GStreamer, DirectShow, OpenGL, etc., all run in the Application Layer.

Signal processing functions typically run on a DSP such as a C64x+ core. Hardware accelerators, generally referred to as AC1, AC2, etc., may implement specific acceleration for certain functions. For example, HDVICP, is a High Definition Video and Image Co-processor.

Codec Engine provided by Texas Instruments abstracts many of the complexities of using the signal processing layer and manages resources such as DMA channels, memory required by the signal processing functions, etc.

BENEFITS OF USING THE SCALABLE TI FRAMEWORK

There are several benefits to using the TI software framework, enumerated below and followed with a brief explanation.

- ☐ Reduce time to market
- ☐ Focus on your area of competence while leveraging the rest of the system from TI
- ☐ Easily replace existing IP with best-in-class components
- ☐ Easily migrate from one type of hardware to another type of hardware platform

REDUCE TIME TO MARKET

Leveraging software from Texas Instruments significantly reduces development time and increases efficiency. Developing drivers, evaluating their performance and integrating with a standard Operating System, such as Linux or WinCE, takes a significant amount of effort. Similarly, developing, optimizing and tuning video quality at the right bit-rate takes another significant amount of time. Figure 4 shows an example of savings in engineering effort. By obtaining the software development kit (DVSDK) from TI, you learn from the sample application and build on top of it.

FOCUS ON YOUR AREA OF COMPETENCE WHILE LEVERAGING THE REST FROM TI

Building a real time embedded system requires deep expertise in several very different areas. For example, to build a real time video and audio system, experts are needed in device drivers, application software, signal processing software, in particular video encoders and decoders and front end signal conditioning. What if the majority of developers differentiate on application software, networking and Graphical User Interface? How can they leverage the benefits of signal processing without having to know the details? By obtaining custom software builds or 'codec combos' for their specific segment either from TI or TI's vast network of 3Ps, developers just need to understand the Application Programming Interface (APIs) to utilize the signal processing functions. This lets them use DSPs as though they are using fixed function ASICs or 'black-box,' yet have the flexibility to customize the signal processing functions later if needed.

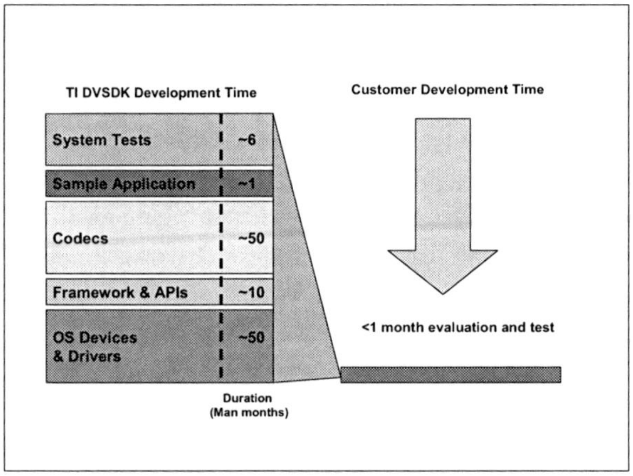

Figure 4: Example saving in engineering effort by leveraging TI platform.

"Codec combos" are specific combinations of encoders and decoders (codecs) targeted for a specific application. For example, in Video Security applications, Figure 5 shows a typical codec combo that consists of an MPEG4 encoder at 720p resolution, combined with another MPEG4 encoder at CIF resolution, combined with another JPEG encoder at D1 resolution with an optional speech codec module.

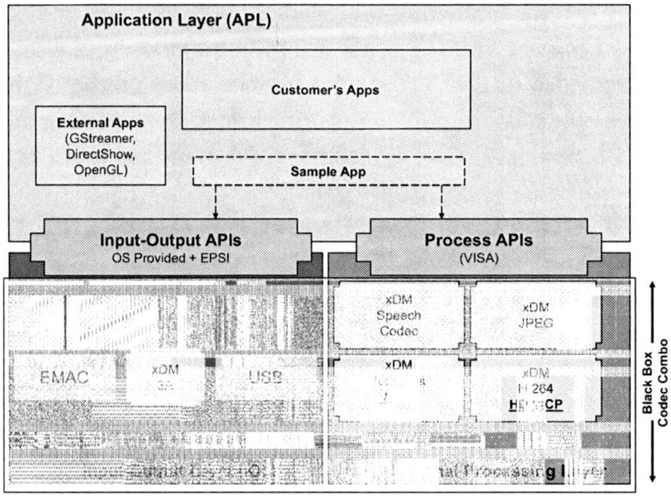

Figure 5: Treat the signal processing layer and IO layer as a black box and focus on your differentiation.

EASILY REPLACE EXISTING IP WITH BEST-IN-CLASS COMPONENTS

After having developed and tested an engineering prototype or product, you realize that one of the components is not up to expectations; or one of your other engineering divisions has developed a better one; or an external 3P is offering a superior component for licensing; or TI has released a newer version with better performance; you need to know if the entire software has to be re-designed and tested at the expense of both engineering cost and time. There may be ripple effects on the rest of the system. The answer is to follow and use the xDAIS for Digital Media (xDM) development guidelines and Application Programming Interface (APIs) from TI. Developers can choose to use their own internal IP that has been made xDM-compliant or pick one from the robust TI technology portfolio (`http://www.thedavincieffect.com/`) or IP from third party (3P) network that support TI devices and are already xDM-compliant. Using xDM, you can replace existing IP without affecting the rest of the system as shown in Figure 6.

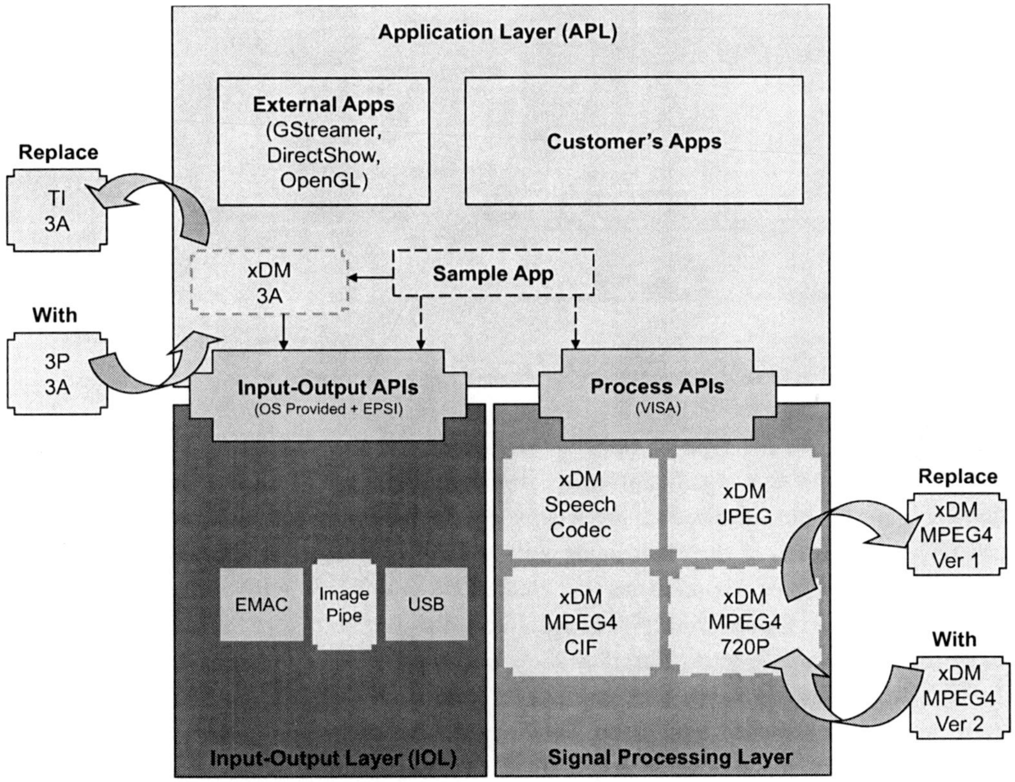

Figure 6: Easily replace xDM components without affecting rest of the system.

EASILY MIGRATE FROM ONE TYPE OF HARDWARE TO ANOTHER TYPE OF HARDWARE PLATFORM

When TI introduces a new silicon platform with a compelling cost-performance mix that opens the door to another segment of the market, how can you reuse your current investment? For example, can you reuse an IPNetcam solution in a Digital Video Recorder (DVR) market since there are several common components? Can you migrate to the next generation hardware platform leveraging the previous generation's effort? You can do all this by consciously using the TI framework and following the guidelines. This is shown in general for any application in Figure 7.

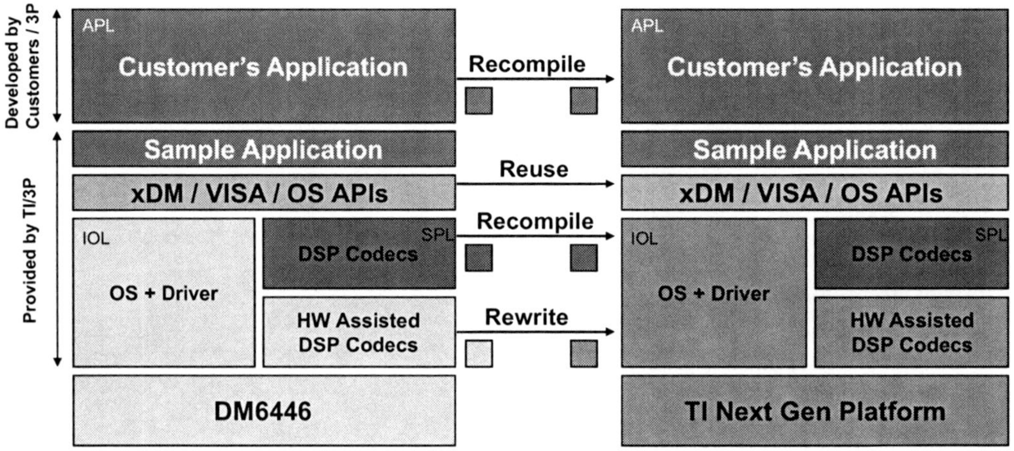

Figure 7: Easily able to migrate software for diverse hardware platform.

Depending on the type of device, the drivers and the Operating System may have to be re-written; however, this significant effort is usually done by TI and/or its rich network of 3Ps. Depending on whether the codecs are leveraging the hardware accelerators, they may or may not have to be re-written. If they run using only a DSP core such as C64x+, they need to be just recompiled. The rest of the software that calls xDM and VISA APIs will usually be recompiled. Figure 8 and Figure 9 show examples of specific end equipments such as an IP security camera and a networked projector; in particular, they show which parts are re-written, recompiled, and re-used. The IPNetcam example shows how one can migrate from DM355, an ARM+Accelerator type of device, to the next generation platform. Similarly, the networked projector example shows how one can migrate from OMAP353x, an ARM+DSP+Accelerator type of device, to the next generation platform.

SCOPE OF TI SOFTWARE (IS – IS NOT)

The scope of this document is limited to the software offered by Texas Instruments on development platforms called DVSDK – Digital Video Software Development Kits. The two main software

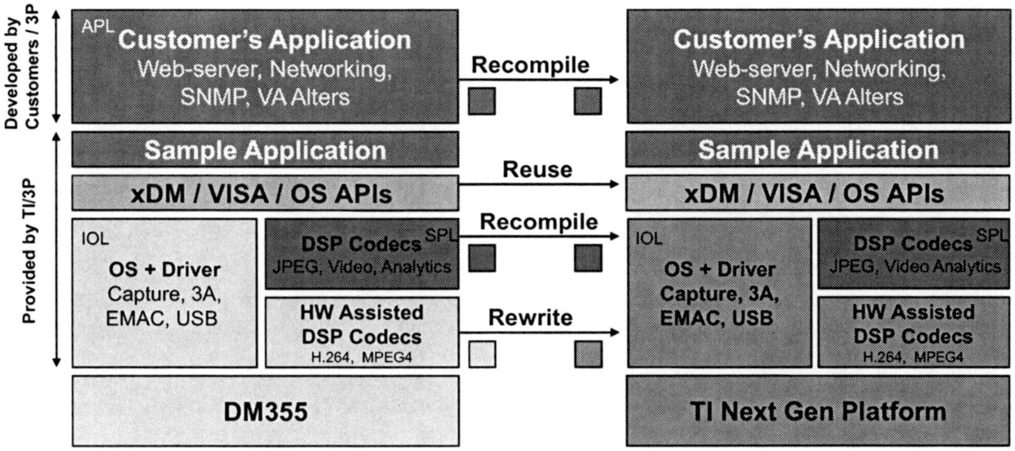

Figure 8: Example: Video security IP network camera.

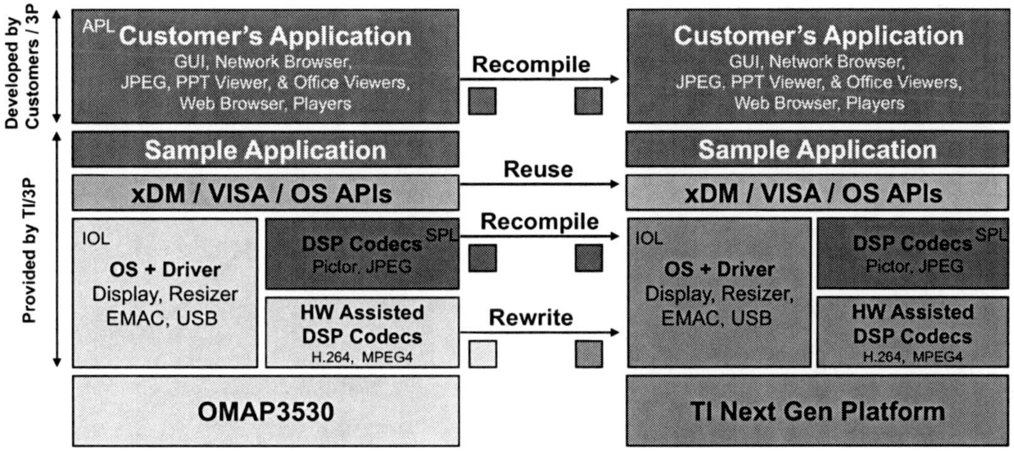

Figure 9: Example: Network projector.

components offered are (1) drivers integrated into an Operating System (OS) such as Linux or WinCE, and (2) signal processing algorithms such as video encoders and decoders. We address how to utilize these two software components, the benefits of TI software framework, and how to migrate from one TI development platform to another.

VISA and xDM APIs are by design OS-agnostic and can readily be used in a variety of higher level software frameworks such as GStreamer, DirectShow, OpenMax, etc. End-equipment or Product-specific application software can directly call VISA or xDM or call

GStreamer/DirectShow/OpenMax/OpenGL/OpenVB APIs, which in turn call VISA or xDM APIs.

The Sample Application addressed in this document is a simple piece of software that calls the TI APIs. It is meant as reference code with the main purpose of teaching or showing how to exercise the TI APIs. The Sample Application should not be confused with product-specific application software, which by itself can be huge.

This document does not address how to develop application level software. This is the domain of our customers' development and engineering teams. Depending on the type of product, the effort involved in developing application software may vary significantly. We categorize Application level software as all software related to user interface, networking stacks, wireline and wireless protocols, web browsers, media players, diagnostics and so on. Some of these may be developed by our customers internally, or they can obtain them from the open source in the case of Linux systems or procure them from third parties. Figure 10 shows the relationship between external APIs such as GStreamer or DirectShow and TI APIs.

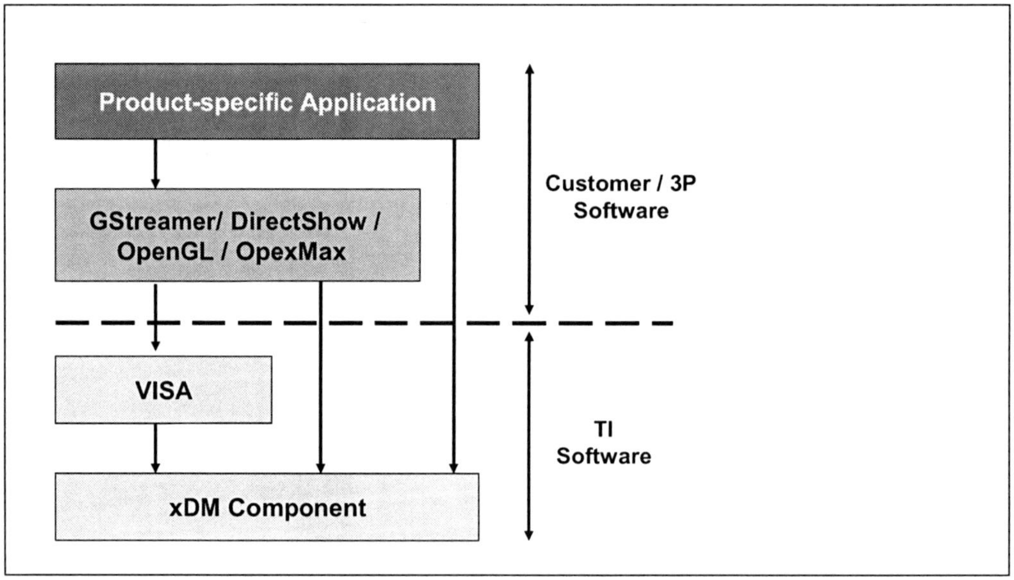

Figure 10: Relationship between external APIs and TI APIs.

SOFTWARE AVAILABLE FROM TEXAS INSTRUMENTS

The following URLs list the software available from Texas Instruments along with procedures for obtaining them.

For overall information on codecs, please go to
www.ti.com/dms
To see additional software available by request, please go to
www.ti.com/requestfreesoftware

CHAPTER 1

Software Platform

1.1 INTRODUCTION

Before Texas Instruments used to offer Digital Signal Processors (DSPs) with an evaluation module, called EVM, bundled with a compiler and debug tools, a software developer had to build his product from the ground up, starting with drivers for accessing the peripherals, to signal processing functions and an application for user interface. But signal processing applications have since evolved from being a few hundred lines of code to several thousand lines of code. The underlying processors have also grown in complexity and capability. Keeping with this trend, Texas Instruments now offers building block software components along with the EVMs. These software components are rigorously tested, documented and released as software products. In addition, they are well supported by a network of external partners or third parties (3Ps). The driving application has evolved from speech and audio to now include video and imaging application that demand high performance, bandwidth and memory. These EVMs prepackaged with component software, and an Operating System are now referred to as Digital Video Software Development kits or DVSDKs. They form a base platform on top of which a developer can build his product.

In the remainder of this chapter, we introduce the types of software products offered by Texas Instruments and a software architecture that shows how a developer can use them to create his application.

While we continue to offer many different types of hardware platforms so that you have a choice and can select the right device that meets your requirements, later in this chapter *we introduce one software platform that is common across these different hardware platforms.*

1.2 THREE SOFTWARE PRODUCTS ALONG WITH APIS

Texas Instruments provides three main software components along with a published set of Application Programmers' Interface (APIs) that show how to call these software components. The three main software components are:

1. Codecs and codec combos
2. Drivers integrated into an Operating System (OS)
3. Domain-specific acceleration libraries

A sample application program shows how to call the codecs/codec-combols and the input-output drivers. Codecs and codec combos are processing functions that take in a buffer of data, process it and output another buffer of data. Drivers integrated into an OS are input-output functions that either capture real time data or display real time data. These two products are meant for application developers and are delivered by either Texas Instruments or its partners.

The third product, domain-specific acceleration libraries are targeted at algorithm developers. These libraries consist of a base set of processing kernels useful in specific domains. As an example, a video analytics library consists of some of the popular processing functions used in object tracking and recognition algorithms. This third category of product is generally not available.
The three APIs are:

1. **xDM**: xDAIS for Digital Media (xDM) API for plugging, playing and easily replacing signal processing algorithms, for example, codecs.

2. **VISA**: Video, Imaging, Speech & Audio (VISA) APIs that abstract the details of signal processing functionality, and allow an application developer to leverage the benefits of these functions without having to know their details.

3. **EPSI**: Embedded Peripheral Software Interface (EPSI) for application developers to input and output data needed by the signal processing functions.

Behind this software architecture and APIs, transparent to the developer, is the **Codec Engine (CE)** that abstracts the complexities of the signal processing functionality and manages the precious system resources for real time processing.

1.2.1 CODECS AND CODEC COMBOS

The term **codec** refers to a component that contains both an encoder and a decoder. While speech codecs contain both an encoder as well as a decoder, video codecs usually do not. The encoder in video codecs is significantly more complex than the decoder. As a result, in video codecs are delivered either as encoders only or decoders. All codecs delivered by TI are **xDM compliant**.

A **codec combo** contains two or more combinations of the same or different codecs. For example, video security applications usually record the same view at two different resolutions; one at a low resolution for real time viewing and another at a higher resolution for later viewing. As a result the same view is encoded with an MPEG4-SP encoder at CIF resolution, as well as MPEG4-SP encoder at D1 resolution. Both encoders need to run in real time at 30 frames per second. In yet another application, an MPEG2 video decoder may need to run simultaneously with a WMA audio decoder.

While a single codec can allocate all the device resources that it needs, such as fast on chip memory, DMA channels, etc., a codec combo may not have this luxury. System resources such as fast on-chip memory, registers, and DMA channels are limited on any type of device. When two or more codecs need to run simultaneously, resources required by the second codec may not be allocatable. In this situation, knowing the real time requirements of the target application, a software integrator can make the *necessary quality versus performance(MHz) tradeoffs* for each individual codec, and allocate just the right amount of on-chip resources needed by each codec and push the remaining to off chip resources. This results in a *resource optimized combination* of codecs that allows the codecs to run simultaneously and meet real time requirements. *Creating a codec combo requires a good understanding of the target application and use-cases so that the right engineering tradeoffs in terms of performance and quality can be made.*

Codecs and codec combos are delivered by Texas Instruments and its rich ecosystem of third party companies (3P).

1.2.2 DRIVERS INTEGRATED INTO AN OS

Drivers are pieces of software that allow an application developer to open, read, write and close the peripherals. Without drivers, you would have to know the details of the registers for each peripheral and program at the bit level. This is obviously time consuming and error prone. Instead of just providing the drivers, we have gone one level higher in abstraction. We provide an Operating System, such as Linux or WinCE, with the drivers already optimized, integrated and tested. This base support package enables the developer to focus on developing the application and not spend time and effort in understanding the mundane management of register bits for each peripheral.

This software product is offered either by Texas Instruments or its third party companies. For example, for the DM6446 device, Monta Vista provides the Linux support package; for OMAP353x, BSquare provides a WinCE support package through Texas Instruments.

1.2.3 DOMAIN-SPECIFIC ACCELERATOR LIBRARIES

These libraries are software optimized libraries that run on a C64x+ core or dedicated hardware specialized for processing basic functions relevant to a specific domain. The DSPLIB, for example, is a C64x+ based library that contains most of the basic functions or kernels needed in the domain of signal processing. Similarly, IMGLIB is an optimized library for the domain of image processing. Some devices like the DM6446, besides having the C64x+ core, also additionally have a hardware assist targeted at video and image processing functions. This hardware assist is called the Video and Imaging Coprocessor (VICP). While DM6446 supports Standard Definition video processing, the DM6467 has a High Definition VICP hardware assist for supporting video processing at high definition resolutions.

Another example of a domain specific kernel library is the Video Analytics (VA) library. It contains basic functions typically needed by Video Analytics algorithms. Providers of VA algorithms may now leverage these libraries to significantly boost the performance of their applications on TI platforms.

Some of these libraries, such as the Image library (IMGLIB) or DSP library (DSPLIB), are provided in source format with no support from Texas Instruments. Developers can use them as starter ware and further modify them to suit their needs. Other libraries, such as VICP are generally available as object code with an API.

1.3 OBTAINING THESE SOFTWARE PRODUCTS

The following URLs list the software available from Texas Instruments along with procedures for obtaining them.
For overall information on codecs, please go to
`www.ti.com/dms`
To see additional software available by request, please go to
`www.ti.com/requestfreesoftware`

1.4 SOFTWARE ARCHITECTURE

Figure 1.1 shows the software architecture. All the processing functions have been grouped into one layer that we refer to as the Signal Processing Layer (SPL). Similarly, we have grouped all the input

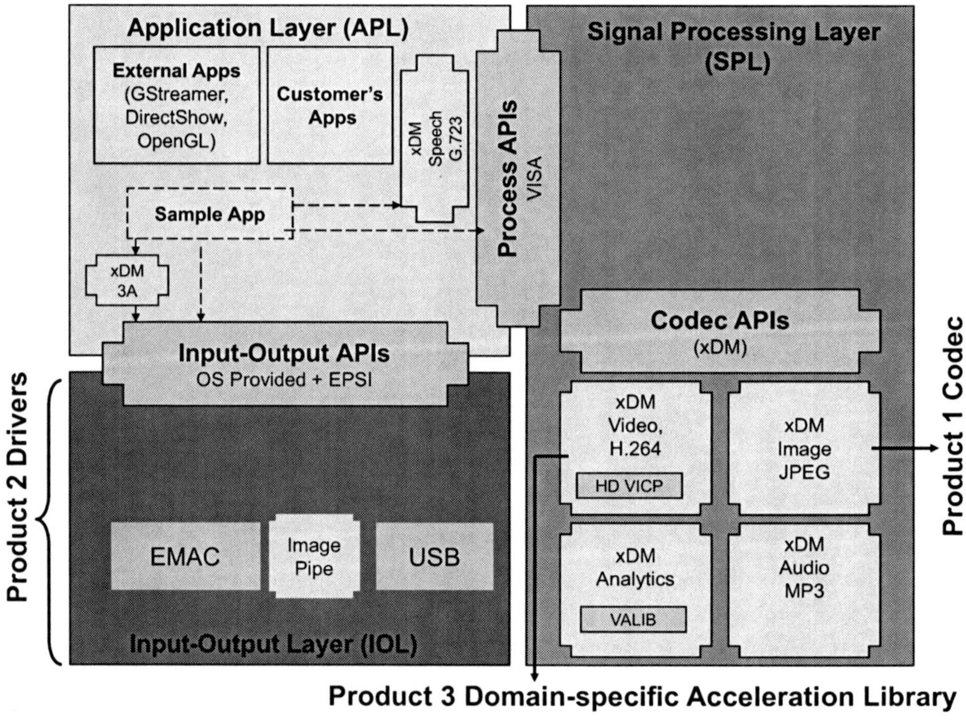

Figure 1.1: Software Architecture Block Diagram.

and output functions into another layer that we call the Input-Output Layer (IOL). The third layer is the Application Layer where we expect you to spend most of your development time. Of course, you may develop components for either the SPL or IOLs.

This software architecture supports the following compute paradigm. A video capture driver, for example, reads data from a video port or peripheral and starts filling a memory buffer. When this input buffer is full, an interrupt is generated by the IOL to the APL and a pointer to this full buffer is passed to the APL. The APL picks up this buffer pointer and in turn generates an interrupt to the SPL and passes the pointer. The SPL now processes the data in this input buffer and when complete, generates an interrupt back to the APL and passes the pointer of the output buffer that it created. The APL passes this output buffer pointer to the IOL commanding it to display it or send it out on the network. Note that only pointers are passed while the buffers remain in place. The overhead passing the pointers is negligible.

1.4.1 SIGNAL PROCESSING LAYER

The Signal Processing Layer (SPL) consists of the entire signal processing functions or algorithms that run on the device. For example, a video codec, such as MPEG4-SP or H.264, will run in this layer. These algorithms are wrapped with xDM API. In between xDM and VISA are the Codec Engine, Link and DSP/BIOS. Memory buffers, along with their pointers, provide input and output to the xDM functions. This decouples the SPL from all other layers. The Signal Processing layer (SPL) presents VISA APIs to all other layers. SPL is delivered as:

- .lib for uniprocessor SoCs such as DM6437.
- .out for multiprocessor SoCs such as DM6446.

1.4.2 INPUT OUTPUT LAYER (IOL)

The Input Output Layer (IOL) covers all the peripheral drivers and generates buffers for inputting or outputting data. Whenever a buffer is full or empty, an interrupt is generated to the APL. Typically, these buffers reside in shared memory, and only pointers are passed from IOL to the APL and eventually to SPL. The IOL is delivered as drivers integrated into an Operating System such as Linux OS or WinCE. In the case of Linux, these drivers reside in the kernel space of Linux OS. The Input Output layer (IOL) presents the OS-provided APIs as well as EPSI APIs to all other layers.

1.4.3 APPLICATION LAYER (APL)

The Application layer interacts with IOL and SPL. It makes calls to IOL for data input and output, and to SPL for processing. The Sample Application Thread (SAT) is a sample application component that shows how to call EPSI and VISA APIS and interfaces with SPL and IOL as built in library functions. All other application components are left to the developer. He may develop them or leverage the vast open source community software. These include, but not limited to, Graphical User Interfaces (GUI), middleware, networking stack, etc.

1.5 SOFTWARE STACK FOR THE DAVINCI & OMAP PLATFORMS OF PROCESSORS

A software stack shows the hierarchical relationship amongst the various software components. Figure 1.2 is a high level software stack. At the bottom, we show a generic hardware platform that

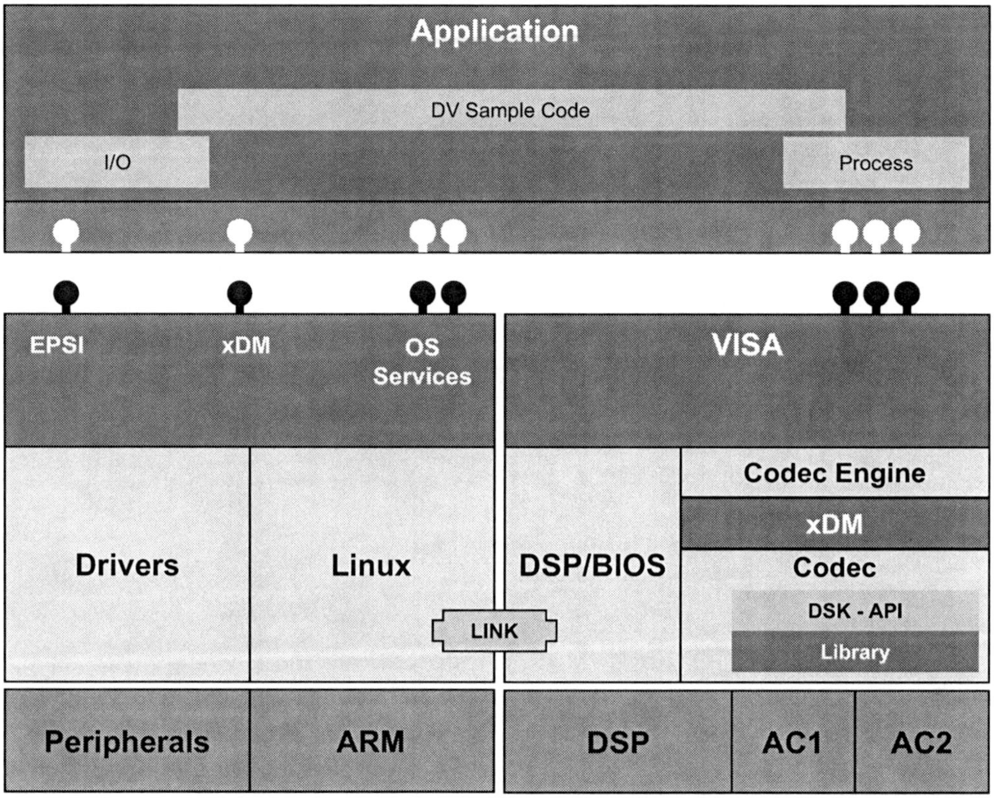

Figure 1.2: Software Stack.

consists of peripherals, a general purpose processing unit such as an ARM-core, a signal processing unit (DSP) such as a C64x+ core, and dedicated hardware assists labeled AC1 and AC2. We can view different hardware platforms as an appropriate combination of these base hardware blocks. For example, the DM6446 device consists of an ARM9 core, a C64x+ core, several peripherals, and one AC1 called Video and Imaging Coprocessor (VICP).

Sitting above the ARM is a standard Operating System such as Linux or WinCE. Figure 1.2 shows the Linux Operating System. Sitting above the DSP is the DSP/BIOS, a real time Operating System provided by Texas Instruments. Sitting above the hardware assist(s) is a library that leverages the hardware assist(s) to provide a basic set of processing kernels that are referred to as Domain-

Specific kernel library (DSK-lib). Standard Definition(SD) and High Definition (HD) Video and Image Co-processors(VICP) SD-VICP and HD-VICP are examples of DSK-lib. Since the DSK-libraries are closest to the hardware, the APIs are different and unique for the domain (video, analytics, imaging, etc.) they accelerate.

In the case of ARM + DSP + Accelerator platform of SoCs, we have arbitrarily assigned the peripherals to ARM/Linux; in reality and from a hardware perspective, the DSP may be able to access the peripherals also. However, this may not be supported in the tested software that TI provides.

DSP/BIOS is an Operating System developed by Texas Instruments. Link is a software product that provides for Inter-processor communication. A portion of this software resides on both the processors that it provides its services to.

An xDM component is a processing function. An example is MPEG4 encoder function. In order for the component to be xDM-compliant, it must adhere to certain implementation rules.

Combined with DSP/BIOS and Link, the Codec Engine (CE) abstracts xDM signal processing functions and presents VISA as the API to the application developer. Some of the services offered by CE include:

- Managing and allocating resources to the xDM algorithm.
- Remote Procedure Calls (RPC) support. An application developer does not have to know where the xDM algorithm is running. For example, the xDM algorithm may be running locally on the DSP, in the SoC or on another discrete DSP attached to the SoC.
- Abstracts and reduces the number of API calls. Instead of making several direct calls to the xDM algorithm, the application has to call/use four APIs: Create, Control, Process, and Delete.
- Insulates the application from changes in the hardware.
- Enables replace ability. It is easy to replace codecs within a class.

In addition to hiding the complexities of the signal processing algorithms, VISA also insulates the application from changes to the signal processing functions. For example, it is easy to replace an MPEG4 codec with a H.264 codec; similarly, it is easy to replace a G.729ab codec with GSM-AMR speech codec since they belong to the same class. *However, changing or replacing the codec has no ripple effect on the application software.*

All peripherals on a device have a driver. These drivers are integrated into the Operating System (OS) and delivered on a hardware platform as part of the OS. Similar to VISA, these drivers also abstract the complexities of the hardware peripherals. Some of these peripherals like the Video Port Sub System (VPSS) are feature-rich and complex. In addition to video capture and display functionality, they provide other dedicated functions like resizing the video frame, histogram calculation, and Onscreen Display(OSD). The basic features of video capture and display are exposed to the developer as V4L2 and FBDEV APIs, which are standard Linux OS APIs. However, there are no equivalent Linux OS APIs for the other hardware features. In this case, TI exposes a TI defined API for this feature.

As shown in Figure 1.3, Embedded Peripheral Software Interface (EPSI) provides *an interface to the drivers that input and output data needed by the signal processing functions.* Note that there are

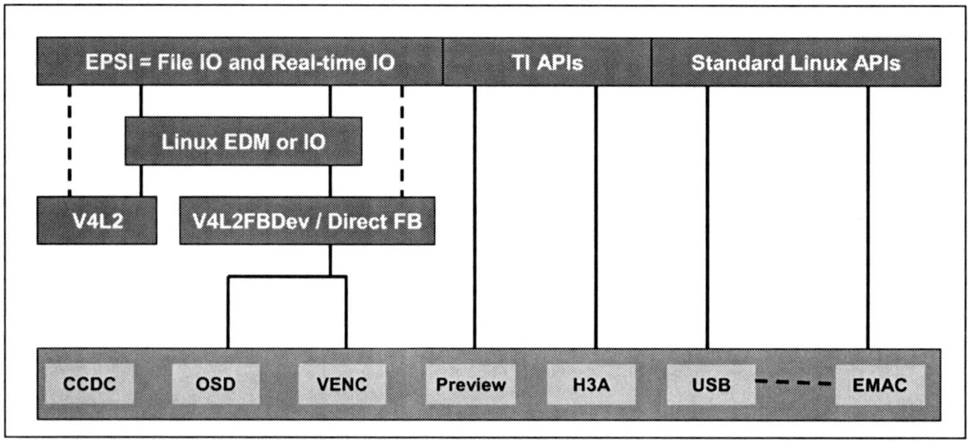

Figure 1.3: Application Software Developers Interface (DM6446 HW Peripherals).

several peripherals on each device and EPSI does not address all the peripherals. EPSI addresses only those services that provide input and output services such as file i/o, and real time video capture and display. The rest of the peripherals have drivers that are exposed to the developer as part of the standard OS services conforming to the OS-provided API.

The goal of EPSI is to maintain a single interface for data input and output irrespective of the OS used. While this does add a small performance overhead, we think the benefit of portability outweighs the overhead penalty. However, for those developers very conscious of overhead, we provide direct access to the driver APIs as well. *Developers now have a choice and can make the appropriate tradeoff between portability and minimum latency.*

1.6 RISING SOFTWARE COSTS & INCREASINGLY DIVERSE HARDWARE PLATFORMS

Texas Instruments continues to offer different hardware devices with varying processing capabilities and peripherals to match application requirements. This rich portfolio of devices can be categorized into three types:

1. DSP + ARM + Accelerator system-on-chip (SoC)

2. DSP only

3. ARM + Accelerator SoC

The first category, **flexible SoCs**, integrate an ARM processor and a DSP core. The second category, **DSP-only devices**, have a single C64x processing core with a rich set of peripheras. The third category, **dedicated SoCs**, contain an ARM processor and a hardware accelerator for a fixed function.

The DM6437 is an example of a DSP-only device optimized for power consumption and cost. It has a C64x+ core and a rich set of peripherals. The second device, the DM648 is a more powerful extension of the DM6437 device. With large on chip memories and an accelerator that runs at different speeds, it is ideally suited for infrastructure applications that require multiple channels of video processing.

The DM6446 is the first member of another type of SoC that integrates an ARM and a fully programmable DSP. This provides an effective solution for products that need flexibility and need to support multiple video formats such as MPEG4 as well as H.264 and JPEG. The ARM processor on the SoC enables application software, networking capability and access to open source code. More recent enhancements to this type of device include the OMAP353x and the DM6467.

The DM355 is the first member of another type of a hardware device. It has an ARM and a dedicated accelerator for MPEG4/JPEG video and image compression. This device provides the most cost effective solution for products that require MPEG4 and JPEG compression functionality. In the future, extensions to this type of a device that implement other dedicated functions may be offered.

While these devices offer abundant choices to a systems developer from a hardware feature perspective, the cost of software development increases from one type of device to another. In addition, since the devices are different at the base hardware level, a software developer's experience in programming these devices is different and varies considerably from one device to another. This in turn has an impact on time to market as well as the engineering effort needed to bring out products. During the next several sections, we will show how we provide a similar software development experience across these different hardware platforms. *This enables our customers to have a single, scalable, and portable software platform and its associated benefits of improved time to market and reduced development costs. Figure 1.4 illustrates this benefit.*

1.7 A SINGLE SOFTWARE INTERFACE ACROSS MANY, DIFFERENT HARDWARE PLATFORMS

Figure 1.5 shows the three types of hardware platforms previously discussed. While the underlying hardware devices are different, we however maintain a single software interface across these different hardware devices. Whether it is the DM6467, or DM6446, or DM355, or OMAP353x device, there is a single, common software platform. This single platform consists of a sample application that shows how to access the peripherals on the device as well the codecs that run on it. The APIs for inputting and outputting data, as well as the APIs for processing the data are held constant across

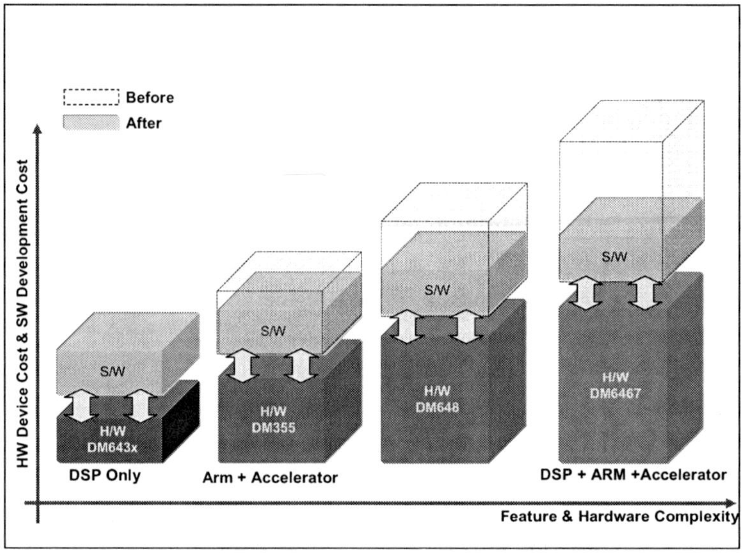

Figure 1.4: Reducing R&D costs for different hardware platforms.

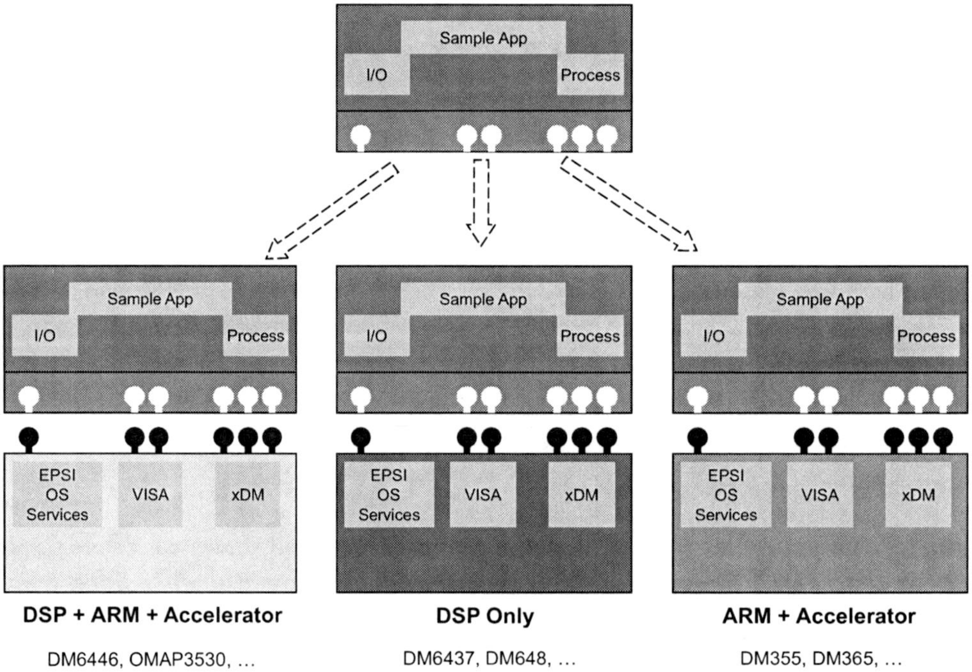

Figure 1.5: One software platform for diverse hardware platforms.

the different platforms for a given OS. As a result, the sample application code at the top is the same while the underlying hardware platforms such as DM6467, DM6446, or DM355 are significantly different.

This single and common software platform consists of:

- **Sample Application software**: This is an example application software that shows how to input, process and output data. It makes use of the three APIs xDM, VISA and EPSI.
- **Base support package**: This is the OS supported on the device with drivers for all the peripherals on the device. The base support package exposes the standard APIs of the OS as well as EPSI.
- **Codecs and codec combos**: This contains the codecs and codec combos supported on the device. xDM and VISA are the APIs to access this functionality.

Figure 1.6 shows the software platform for DSP/BIOS, an Operating System provided by Texas Instruments.

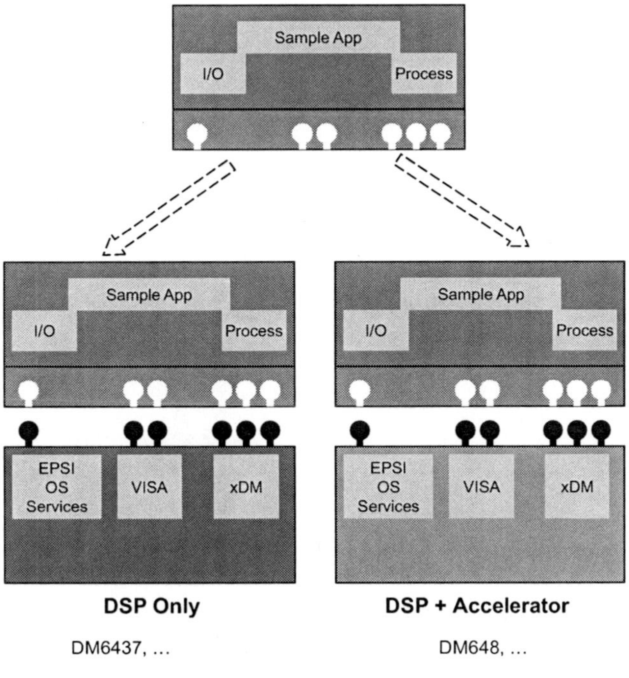

Figure 1.6: Common API set for DSP/BIOS.

CHAPTER 2

More about xDM, VISA, & CE

2.1 DIFFERENT LEVELS OF ABSTRACTION – XDM, VISA AND CODEC ENGINE(CE)

Signal processing functions have attained the perception of being math-intensive, complex systems restricted to a limited set of dedicated engineering developers. Our goal with the Signal Processing Layer (SPL) was to dispel this myth and create a level of abstraction that would enable a large majority of software developers to utilize the benefits of signal processing without necessarily having to understand the details. In the sections that follow, we discuss how this abstraction is provided by xDM, Codec Engine and VISA. xDM, which stands for xDAIS for Digital Media is a standard API that is wrapped around all signal processing functions, especially codecs. VISA APIs explained below are the APIs presented to the application developer. Codec Engine is a piece of software, developed by Texas Instruments, that manages the system resources and translates VISA calls into xDM calls.

2.1.1 XDM COMPLIANT SOFTWARE COMPONENT

In order to view any signal processing function as a black-box, we created a standard called xDM, which stands for xDAIS for Digital Media. xDM components may run on either the DSP or the ARM processor. Since xDM components form the basis of our software framework, we have dedicated an entire chapter to it later in this book.

xDAIS is a standard algorithm wrapper and set of guidelines for creating a real time software algorithm or function. This was developed by Texas Instruments in the late 1990s. While xDAIS was a good, first step towards creating real time software components, we found that every algorithm tended to have its own unique set of APIs. However, when there are several algorithms that need to run in a system, the system integrator and application developer need to learn and become familiar with all the APIs of the different functions. This is, naturally, not an easy task.

In Figure 2.1, we show the xDM APIs categorized into

- Resource allocation, initialization and start APIs.
- Runtime process & control APIs.
- Stop algorithm and resource de-allocation APIs.

The xDM algorithm may in turn exploit the hardware accelerators if available in the SoC. For example, in DM6446 SoC, there is only one hardware accelerator, called Video & Imaging Coprocessor (VICP). An internal set of APIs, called SD-VICP API, provides an abstraction to the

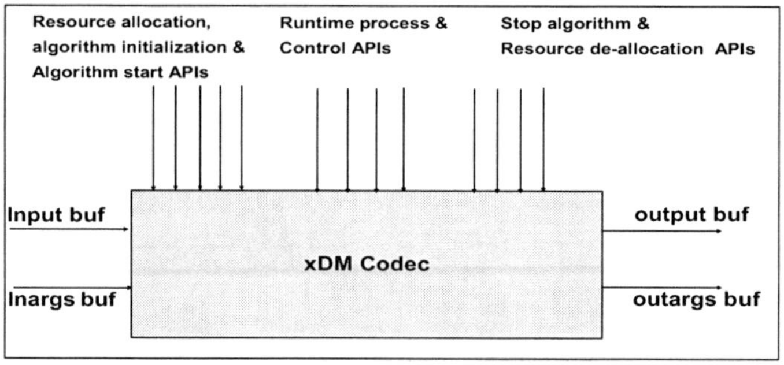

Figure 2.1: APIs that constitute xDM.

xDM algorithm that may choose to exploit it for implementing a specific algorithm such as H.264 video codec. In another device, such as DM6467x, there are two dedicated accelerators for High Definition (HD) video processing called AC1 and AC2. HD-VICP APIs abstracts the complexities of these accelerators and provides an interface to the set of processing functions that run on them.

2.1.2 CATEGORIZING CODECS INTO CLASSES

When we surveyed codecs that are prevalent, we were surprised to find more than sixty different ones. These span different domains, from speech to audio, from imaging to video applications. They were developed by different standard bodies at different times, and naturally have separate APIs. It is a daunting task to learn and use these codecs easily within a system.

As a first step, we categorized these sixty or so codecs into four major categories or classes, namely Video, Imaging, Speech and Audio. This is shown in Figure 2.2. We then created an API for each class. We also tried to maintain the same look and feel across the four classes. This lead to the creation of four, basic xDM classes.

While we initially focused on codecs and the four classes, since then other new classes have been introduced. For example, we worked with several algorithm providers in the domain of Video Analytics (VA) and created a draft proposal of a new, fifth class. Similarly, we introduced a new class for a pre-processing function called 3A which stands for Auto-exposure(AE), Auto-white balance (AWB) and Auto-focus algorithms. In the future, we will continue to work with our partners, developers and customers, and when appropriate, add new classes. Of course, we do not wish to add too many new classes since it would defeat the very purpose of why xDM was created in the first place.

2.1.3 BENEFITS OF XDM

xDM is the foundation of the software platform. It provides a consistent interface to a *class of algorithms*. It has been designed to be both Operating System as well as framework agnostic. As

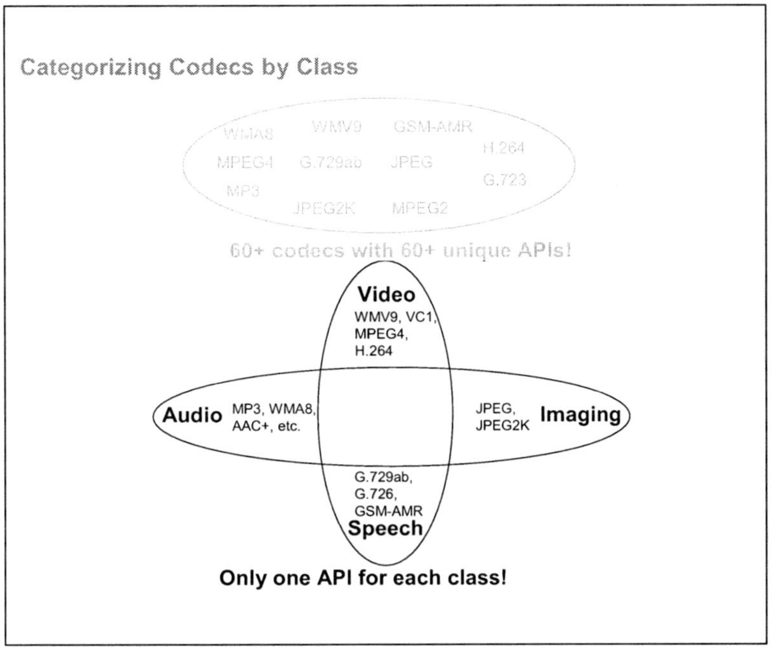

Figure 2.2: Classes with only 1 API for each class.

such, it is easily supported in Linux, WinCE or any other Operating Systems. By design, it has been developed to enable replace-ability and insulate the application code from any ripple effects.

2.1.4 CODEC ENGINE & VISA

While xDM provides a consistent interface across signal processing functions, if an application developer were to use it, he would have to allocate DMA channels and manage other system resources. In addition, if he were developing software on a SoC with two processors, he would also have to learn or develop the inter processor communications (IPC) protocol. This would take away the focus from just using an algorithm in an application, and he would now have to dive deep into the system software. This is where the Codec Engine(CE) comes into picture to manage the system resources and functions. All these lower level management and control functions are now handled by the Codec Engine which manages the xDM component and abstracts the application developer from the signal processing layer.

CE presents a simple and consistent set of interfaces to the application developer called VISA, which stands for Video, Imaging, Speech and Audio. Each of these four groups supports encoders and decoders and the APIs that support them are:

1. VIDENC – video encode

2. VIDDEC – video decode
3. IMGENC – image encode
4. IMGDEC – image decode
5. SPHENC – speech encode
6. SPHDEC – speech decode
7. AUDENC – audio encode
8. AUDDEC – audio decode

Within each set, mentioned above, there are four APIs:

1. `xxx_create()`
2. `xxx_control()`
3. `xxx_process()`
4. `xxx_delete()`

The codec engine software basically translates these *create, control, process and delete* APIs to their respective xDM APIs, while managing the system resources and inter-processor communication. This is shown in Figure 2.3.

create() API creates an instance of xDM algorithm and allocates the required resources for the algorithm to run. The SPL has a certain amount of resources such as DMA channels, on chip memory, cache, etc. Create() API, using the Codec Engine, queries the xDM algorithm for the resources that it needs, and based on the algorithm's requirements, it allocates them. Note that xDM-compliant functions cannot allocate resources directly; they can only request for the resources; the Codec Engine is always in control of the resources and manages them across multiple functions running in SPL.

control() API allows the APL to modify parameters that control the algorithm. Every algorithm exports some parameters that allow its behavior to be controlled. For example, in a MPEG4 video codec, you may wish to change the bit-rate or resolution. Both these arguments would be exported as bit-rate and resolution parameters and the APL using the control() API will be able to change them. Control() API allows the parameters to be changed dynamically from one frame to another.

process () API filters the input buffer to get the output buffer e.g., encode or decode function. For example, an MPEG4 algorithm would use the input buffer, encode it and create an encoded frame in an output buffer.

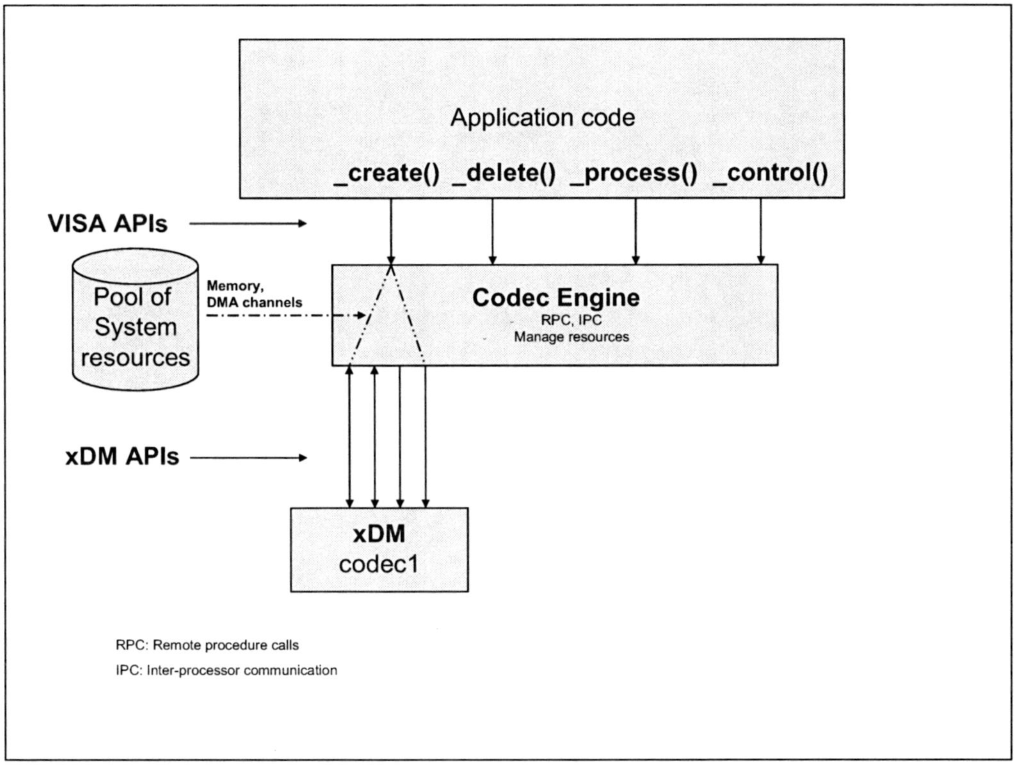

Figure 2.3: VISA.

delete() API deletes the algorithm instance and reclaims the resources. delete() API is basically the complement of create() API. Once the APL decides to stop the xDM algorithm, it calls this API, which reclaims the resources. Then, these are available to another xDM algorithm when required.

The process and control API of VISA are a direct reflection of the low-level process and control functions of the xDM algorithm. *As a result, we're providing low-level control of codecs along with high level abstraction of the details.* In Figure 2.4, we show the specific VISA and xDM APIs. The sequence of VISA API and the hidden corresponding codec engine calls to xDM algorithm is shown in Figure 6.1.

The APL developer needs to understand only these four APIs. For a given class, say Video (or Imaging, or Speech or Audio), the signature of these APIs is held constant. This enables an integrator and developer to easily replace one codec with another.

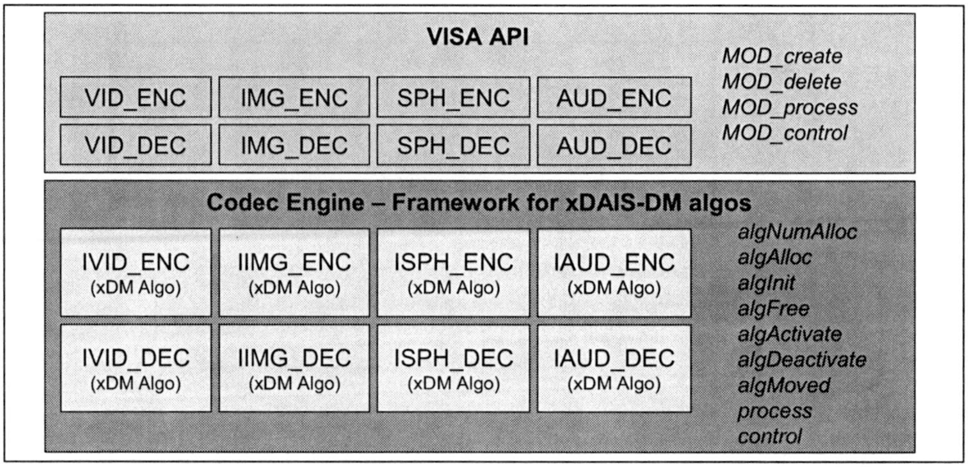

Figure 2.4: VISA Abstracts Details of xDM Algos.

2.1.5 BENEFITS OF VISA API

VISA provides several benefits. We present them from the perspectives of an application developer/author as well as an xDM component provider.

An application author enjoys all the benefits of signal processing layer without necessarily understanding the complexities of the underlying DSP algorithm or hardware. While several codecs are supported, he need only learn and use one API for a given media engine class. Changing a codec within the class involves no changes to application level code. All media engine classes have a similar look and feel.

An application developer can quickly evaluate different implementations of an algorithm sourced by different vendors. This enables him to provide for a *best in class functionality and make the necessary tradeoff between quality, performance and cost.*

In addition, an application developer can leverage other multimedia frameworks such as gstreamer, which can be built on top of VISA APIs.

A component provider or author can conform to a well defined interface. *He need not necessarily understand or have complete knowledge of the end application where the component provides service. A single codec may support several different applications.*

Table 2.1: APIs Definition and Benefits

API	Definition	Benefits
xDM	Std API wrapped along all signal processing algos	Single, consistent interface for all algorithms within a class Easily replace algorithms within a class Insulate system from changes in algorithm
VISA	Std API provided at APL for accessing signal processing functions	Abstracts application developer from signal processing complexities Easily replace algorithms within a class Insulate system from changes in SPL APL does not have to know where the function is executing, i.e., locally or remotely
EPSI	Collection of industry standard APIs and TI defined APIs for peripherals	abstracts application developer from complexities and details of peripheral management Insulates application from evolution in peripheral hardware

CHAPTER 3

Building a Product/Application Based on DaVinci Technology—An Example

3.1 CREATION OF AN INTERNET PROTOCOL (IP) BASED NETWORK CAMERA (IPNETCAM)

In this section, we will show how DaVinci technology is used in the creation of an Internet Protocol (IP) based network camera (IPNetcam). In addition, we show how the same software framework is used on two different hardware platforms, the DM6446 and the DM355 and how they can migrate to DMNext, when such a device is introduced.

Figure 3.1 shows the processing blocks of an IPNetcam. Data captured by a front end sensor is

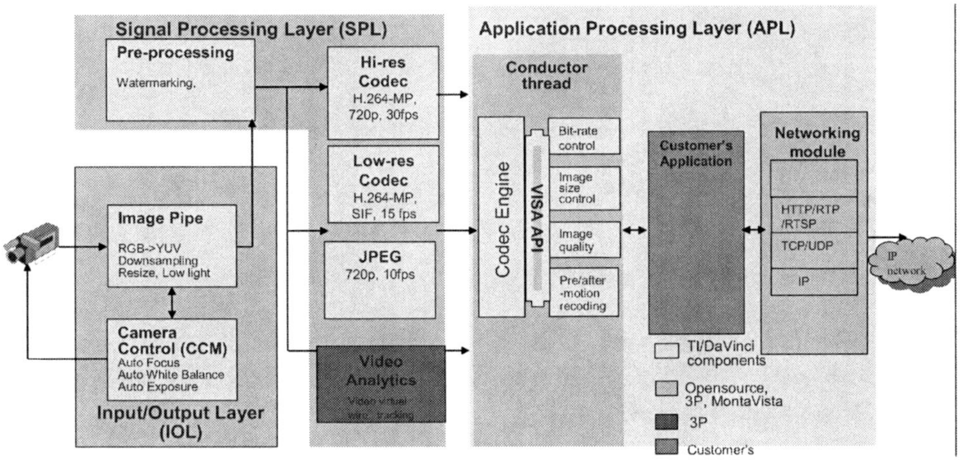

Figure 3.1: Data flow and processing block diagram of an IPNetcam.

input to the IOL that creates a YUV-stream after correcting for bad pixels, color, lens distortion etc. This front end processing is collectively referred to as the Image Pipe. Depending on the input source, the data stream may have to be de-interlaced and a time-stamp (optional) applied to each frame for security reasons. The YUV-frames are then input to a High resolution(D1, 720p or higher) encoder as well as to a low-latency, lower resolution (SIF) encoder. The encoder is typically an

MPEG4 encoder (the industry is now transitioning to H.264 encoder). The encoded stream is then packetized and using either RTP or RTSP, the stream is put out on the network. In addition, there is a significant amount of application software that enables the camera to be a web-server.

From a product perspective, there are three different possible product lines. One is a MPEG4-based product that leverages the DM355 SoC. The other is a H.264-based product combined with Video-Analytics. This second product line can support MPEG4 as well. This second product line is based on the DM6446 SoC and the increased flexibility comes at increased cost. A next generation SoC, DMNext, might be more cost optimized while yet supporting H.264. These different product lines are shown in Figure 3.2.

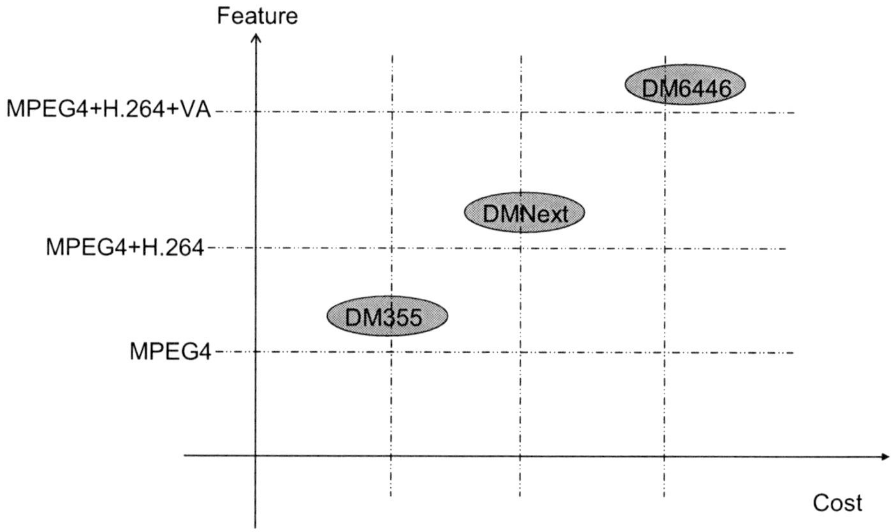

Figure 3.2: Feature versus cost tradeoff of three different IPNetcams.

If a product vendor supports three different product lines that are based on three different SoCs, he can migrate to DMNext when it becomes available yet leverage all the software investment that he has made on the first product line.

He can do this by adopting the TI-provided APIs, xDM, VISA and EPSI and building his application software on top of TI's framework. If his focus is on the application layer, he would leverage these APIs and create his software in the APL. Some vendors prefer to differentiate at the SPL by developing their own unique encoder. They can do so by implementing their encoder and conforming to xDM APIs. By doing so, they will be able to leverage the rest of the software components provided from TI and 3Ps. As shown in Figure 3.3, developers need only change the codec processing module that calls the VISA APIs, and input-output processing module, that calls the EPSI APIs. Most of these changes relate to specifying which codec to use and the bit-rates for each and the number of instances that need to be created.

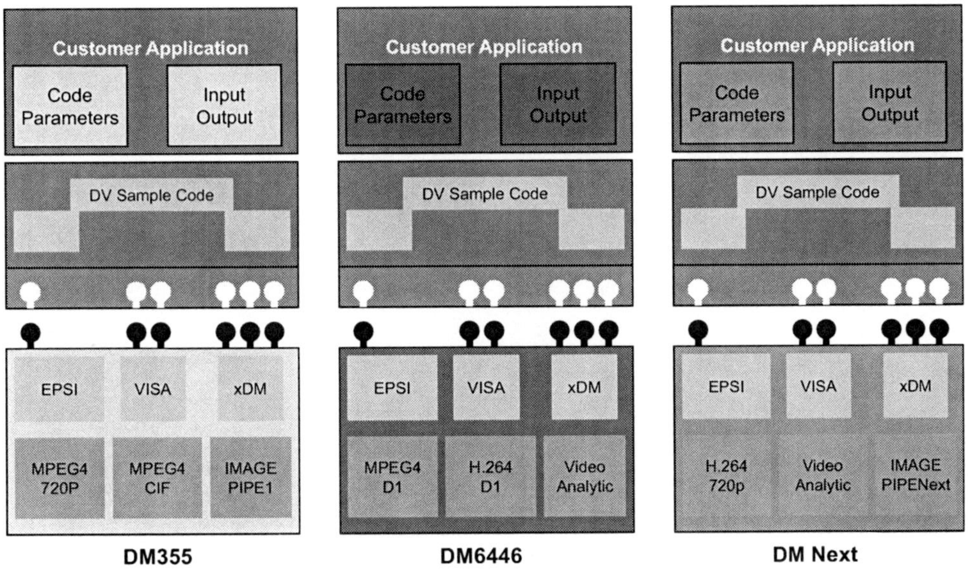

Figure 3.3: Changes in APL as developer migrates from one platform to another.

Chapter 9 shows an example application written on top of this scalable software platform. This example application is referred to as the Digital Video Test Bench (DVTB) and is supplied with the DVSDK.

CHAPTER 4

Reducing Development Cost While Introducing Multiple Products

4.1 REDUCING DEVELOPMENT COST WHILE INTRODUCING MULTIPLE PRODUCTS

As we discussed earlier, Texas Instruments introduces technological innovations at a rapid pace. You can pick the right device or system-on-chip (SoC) from a portfolio of products. Following the xDM guidelines and using the APIs and software platform from TI enables you to introduce multiple products. For example, you can first introduce a medium-featured product using one type of hardware platform and follow that with a low end product leveraging a totally different hardware platform while significantly reducing the engineering effort in migrating from one to the other hardware platform. In the remainder of this section, we discuss how this is possible and walk through a scenario of starting with a mid-end product and creating a low-end as well as a high-end product.

Suppose you choose the OMAP353x as you first hardware platform. During the development of your mid-end featured product, you will go through several engineering and development steps such as:

1. **Learn and evaluate**: In this step, you will become familiar with the TI development platform, learn about the tool chains, the development environment and the sample application. In addition, you will evaluate the performance of software/hardware platform or DVSDK. For example, you will assess the bandwith or throughput of the drivers, and the MHz consumed by the video codecs or other functions. You will assess the quality of the video codecs by testing them on your test streams and decide how well the performance and quality of the platform meets your requirements. Once you are convinced that the platform meets your requirements, you are ready to proceed to the next step.

2. **Build and prototype**: In this second step, you start building your target application. In addition to the software components from Texas Instruments, you may have your own differentiating algorithms or Intellectual Property (IP). Following the guidelines of xDM, you will create an xDM-compliant algorithm and integrate your IP with the rest of the platform. You then develop an application that runs in the application layer that calls your IP as well as the other software functions provided in the DVSDK using the sample application as a reference.

3. **System Test and Release**: In this third and final step, you rigorously test your product for conforming to your product's use-cases and then release it to the market.

Once this mid-end product is launched, you decide that you need to introduce a low-end product with a subset of the features. For this you choose a lower-cost device from Texas Instruments, say for example, the DM355. This device is quite different from the earlier OMAP353x. However, it has APIs that are similar and the software platform on this device has the same look and feel. You will now go through the same three steps outlined above. However, the amount of engineering effort and time spent in the first step of **Learn and Evaluate** is reduced to one of **Validate and Confirm**. In this step, since you are already familiar with the TI platform, you will just validate the performance of the drivers and codecs on DM355. After confirming that this platform meets your performance expectations, you then proceed to the second step of **Build and Prototype**, which is now reduced to Reuse and Defeature. Here you *re-use your previous build* and de-feature it to create the lower-end product. Depending on how well your application code is structured, you just recompile it for DM355. *Your application code remains unchanged since the same VISA and xDM APIs that were previously available on the OMAP353x are now available on the DM355.* While the underlying video codecs maybe different from an implementation point of view since they now run on a hardware accelerator instead of a programmable DSP, *they still expose the same APIs. As a result, the Application code that called the video codecs in OMAP353x remains unchanged when used on the DM355.* These steps are shown in Figure 4.1.

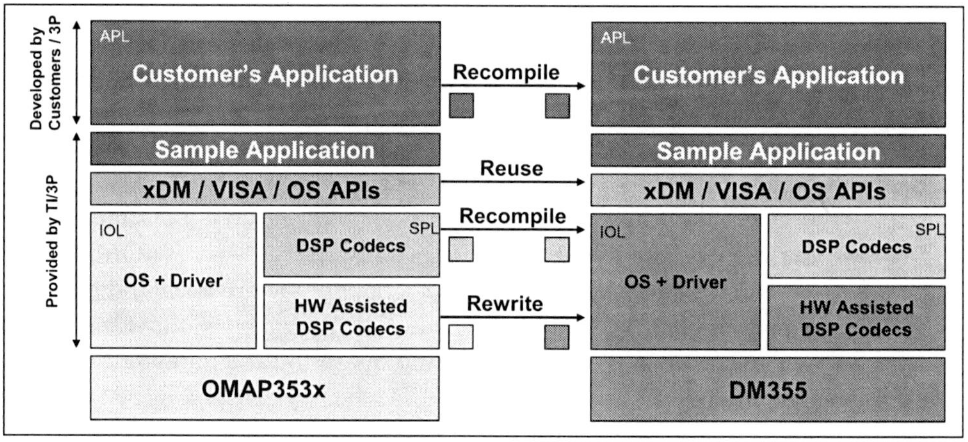

Figure 4.1: Migrating from one type of hardware platform to another.

Similarly, when you start creating a high-end product, you choose the next generation OMAP353x or DM355 device. You receive the software platform from TI/3P and validate the performance of its drivers and codecs. This is the first step of Validate and Confirm. You now wish to add another differentiating or new feature with a new IP, you can follow the xDM guidelines and create an xDM component. Therefore, you can use either the VISA API if it belongs to the appropriate class of a codec or extend VISA to create a new class. If you use the same class of API, then you don't have to do anything to the Codec Engine; however if you decide to create a new class, then you have to develop software components called stubs and skeletons and add them to the Codec Engine framework. As a result, you are encouraged to use the supported class of VISA on the platform. You then integrate and build a new executable using your new IP along with the rest of the software platform. Using the sample code as a reference you will call and test your new IP along with the rest of the system. These tasks all belong to the second step of Reuse and Add. Finally, you conclude with the third step of Test and Release. Figure 4.2 shows the R&D cost in creating these different products.

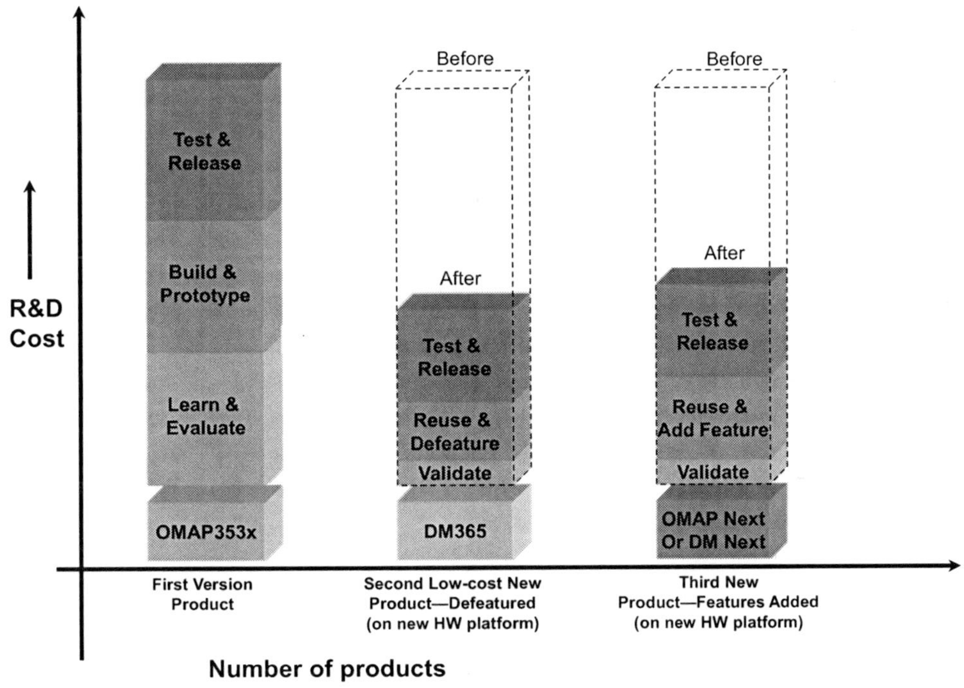

Figure 4.2: R&D cost in creating multiple products from a common codebase.

4.2 SUMMARY

By providing pre-tested, robust software components and a set of well defined APIs and software architecture, we significantly reduce your time to market with new products. The TI development platform consists of codecs, drivers integrated into an OS, Codec Engine, along with APIs, EPSI,

VISA and xDM. An independent test team rigorously tests these components on the DVSDK before releasing them to the market.

At present, xDM addresses four classes of signal processing algorithms in the domains of video, imaging, speech and audio. In addition to these four classes, we are addressing other classes ranging from pre-processing class such as de-interlacing, Auto-Exposure(AE), Auto-White Balance (AWB) algorithms to object recognition class such as video analytics algorithms. xDM APIs for these two new classes will be published when available.

The three APIs, xDM, VISA and EPSI form the foundation APIs that TI strives to maintain consistency from one platform to another. This results in an almost similar software developer experience across diverse hardware platforms. Customers can select from the rich portfolio of TI platforms ranging from DSP-only to ARM+DSP+Accelerator SoCs depending on their product's cost and feature goals. They can first deploy a low-end product with limited features using a low-cost device; later, they can introduce a high-end/mid-end product using a more powerful device while reusing most of their software. In effect, customers can build once for a specific product and deploy many products while preserving most of their software investment.

Technology continues to evolve at such a rapid pace that it is becoming extremely challenging for one vendor to be an expert in all areas needed to create a compelling product. For example, creating an IP Network Camera requires a good front end algorithm, frequently referred to as Image Pipe, for cleaning the signal output by a sensor, in addition to having an optimized video encoder tailored for video security use cases, followed by an application that includes graphical user interfaces, networking and encryption software components. In addition to these basic features, differentiating functionalities such as Video Analytics are also emerging. Conforming and building on top of xDM, VISA and EPSI APIs enables customers to play in their area of expertise while continuing to leverage components provided by different intellectual property (IP) suppliers/3Ps besides Texas Instruments. Developers can easily migrate to newer platforms when Texas Instruments introduces them while preserving their investment on previous platforms.

CHAPTER 5

eXpressDSP Digital Media (xDM)

This chapter was written with Aravindhan K., Member Group Technical Staff, Texas Instruments India Pvt. Ltd.

5.1 INTRODUCTION

The xDM standard defines a uniform set of APIs for multimedia compression algorithms (codecs) with the main intent of providing ease of replaceability and insulate the application from component level changes. xDM is built over TI's well proven eXpress DSP Algorithm Interoperability Standard (also known as xDAIS) specification. This chapter introduces xDAIS and xDM.

In this chapter, we will discuss a portion of the signal-processing layer, in particular, eXpress-DSP Digital Media (xDM) standard. Two main benefits of xDM are

- Replacability - provides the flexibility to use any algorithm without changing the client application code. For example, if you have developed a client application using an xDM-compliant MPEG4 video decoder, then you can easily replace MPEG4 with another xDM-compliant video decoder, for example H.264, with minimal changes to the client application.
- Insulation of the application layer from changes in the components in the signal-processing layer. Any changes to components in the signal processing layer does not result in changes to the application layer.

5.2 OVERVIEW OF XDAIS AND XDM

TI's multimedia codec implementations are based on the eXpressDSP Digital Media (xDM) standard. xDM is an extension of the eXpressDSP Algorithm Interface Standard (xDAIS).

5.2.1 XDAIS OVERVIEW

An eXpressDSP-compliant algorithm is a module that implements the abstract interface IALG. The IALG API takes the memory management function away from the algorithm and places it in the hosting framework. Thus, an interaction occurs between the algorithm and the framework. This interaction allows the client application to allocate memory for the algorithm and share memory between algorithms. It also allows the memory to be moved around while an algorithm is operating

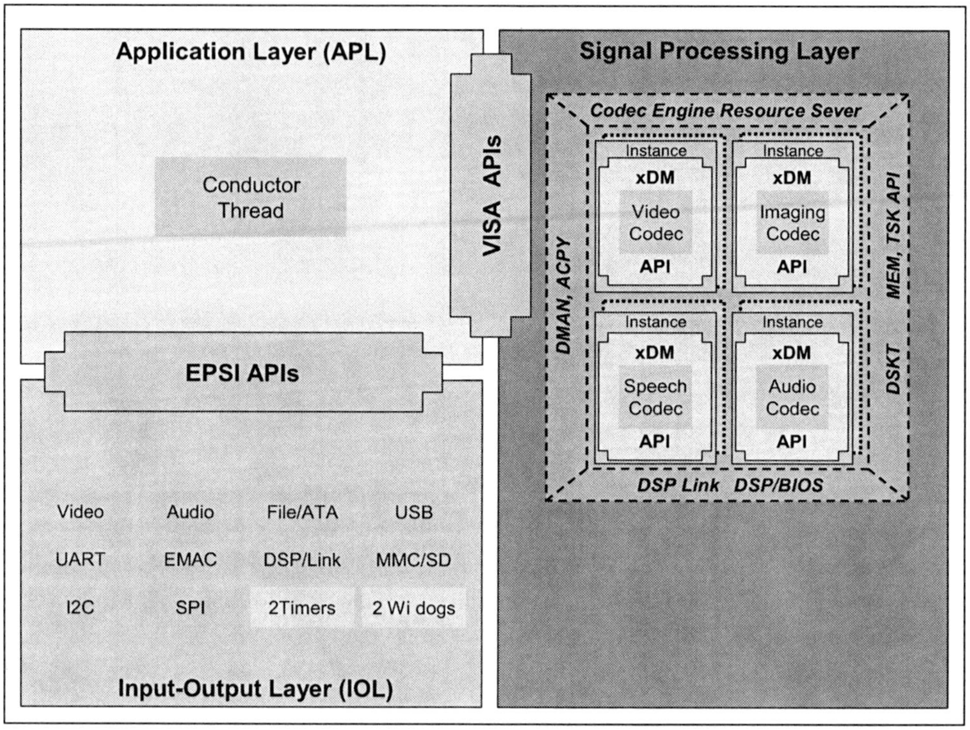

Figure 5.1: Replacability and Insulation.

in the system. In order to facilitate these functionalities, the IALG interface defines the following APIs:

- `algAlloc()`
- `algInit()`
- `algActivate()`
- `algDeactivate()`
- `algFree()`

The `algAlloc()` API allows the algorithm to communicate its memory requirements to the client application. The `algInit()` API allows the algorithm to initialize the memory allocated by the client application. The `algFree()` API allows the algorithm to communicate the memory to be freed when an instance is no longer required.

Once an algorithm instance object is created, it can be used to process data in real-time. The `algActivate()` API provides a notification to the algorithm instance that one or more algorithm

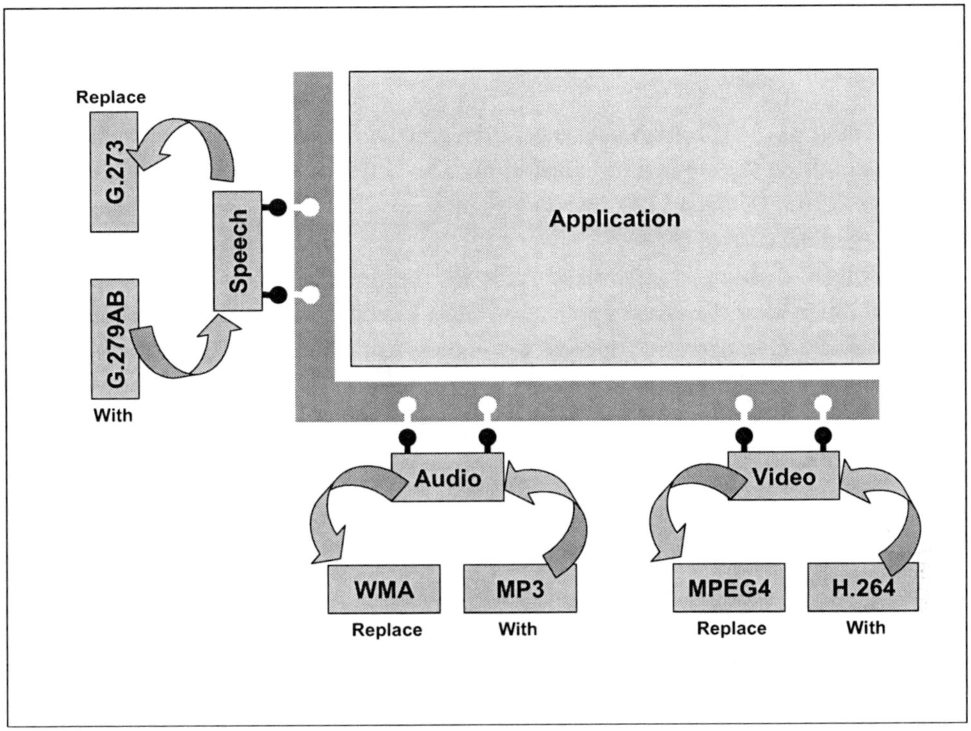

Figure 5.2: xDM enables components to be easily replaced and insulates application.

processing methods is about to be run zero or more times in succession. After the processing methods have been run, the client application calls the `algDeactivate()` API prior to reusing any of the instance's scratch memory.

The IALG interface also defines three more optional APIs `algControl()`, `algNumAlloc()`, and `algMoved()`. For more details on these APIs, see *TMS320 DSP Algorithm Standard API Reference (literature number SPRU360)*.

5.2.2 XDM OVERVIEW

In the multimedia application space, you have the choice of integrating any codec into your multimedia system. For example, if you are building a video decoder system, you can use any of the available video decoders (such as MPEG4, H.263, or H.264) in your system. To enable easy integration with the client application, it is important that all codecs with similar functionality use similar APIs. xDM was primarily defined as an extension to xDAIS to ensure uniformity across different classes of codecs (for example audio, video, image, and speech). The xDM standard defines the following two APIs:

- `control()`
- `process()`

The `control()` API provides a standard way to control an algorithm instance and receive status information from the algorithm in real-time. The `control()` API replaces the `algControl()` API defined as part of the IALG interface. The `process()` API does the basic processing (encode/decode) of data.

Apart from defining standardized APIs for multimedia codecs, xDM also standardizes the generic parameters that the client application must pass to these APIs. The client application can define additional implementation specific parameters using extended data structures.

As depicted in the Figure 5.3, xDM is an extension to xDAIS and forms an interface between the client application and the codec component. xDM insulates the client application from

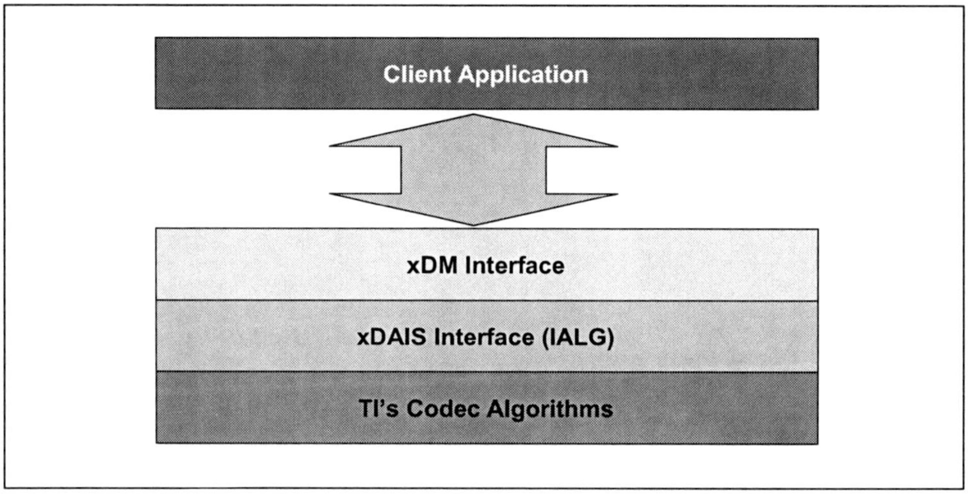

Figure 5.3: xDM interface to the client application.

component-level changes. Since TI's multimedia algorithms are xDM-compliant, it provides you with the flexibility to use any TI algorithm without changing the client application code. For example, if you have developed a client application using an xDM-compliant MPEG4 video decoder, then you can easily replace MPEG4 with another xDM-compliant video decoder, for instance H.264, with minimal changes to the client application.

5.2.3 RELATIONSHIP BETWEEN XDM AND XDAIS-8 CLASSES OF GENERIC INTERFACES

xDM is a superset of xDAIS. One of the driving focus for xDAIS was on providing a standardized interface, IALG, for managing the memory resources needed by an algorithm. The IMOD interface of xDAIS was basically left open to the algorithm provider/developer. This led to a proliferation of custom interfaces unique to each algorithm or codec. As a consequence, it was not easy to *replace one* version of a codec with another slightly different version of the same codec let alone the same codec developed and provided by another vendor. Towards enabling replacability, xDM was defined. Although xDM not only defines a uniform set of APIs, it also specifies the parameters in detail. As a result, all codecs belonging to the same class, such as video, have identical APIs and parameters. Therefore, a video class API may differ from another class such as audio. xDM enables replacability or plug and play within the same class.

xDM defines eight generic interfaces for the following categories. The "x" suffix represents a version of the interface. In xDM 0.9, the suffix was omitted; in xDM 1.0, it is "1."

- ☐ `IVIDENCx` - Generic interface for video encoders
- ☐ `IVIDDECx` - Generic interface for video decoders
- ☐ `IAUDENCx` - Generic interface for audio encoders
- ☐ `IAUDDECx` - Generic interface for audio decoders
- ☐ `ISPHENCx` - Generic interface for speech encoders
- ☐ `ISPHDECx` - Generic interface for speech decoders
- ☐ `IIMGENCx` - Generic interface for image encoders
- ☐ `IIMGDECx` - Generic interface for image decoders

5.2.4 SCOPE OF THE STANDARD

xDM addresses the following:

- ☐ Uniform lightweight APIs across various classes of multimedia algorithms, such as audio, video, speech, and image
- ☐ Flexibility of extension for various requirements such as metadata parsing, file format, custom processing, and so forth
- ☐ Interoperability across various algorithms and vendors

xDM does not address the following:

- ☐ Metadata parsing from multimedia streams
- ☐ File format or multiplex support
- ☐ Digital Rights Managements (DRM) interaction with codecs
- ☐ Call back from algorithms and applications to enable data movement and processing

- ☐ APIs other than codecs, for example, pre- and post-processing APIs like resizing, echo cancellations, and so forth

5.2.5 GOALS OF THE STANDARD

The goals of this standard include:

- ☐ Enable plug and play architecture for multimedia codecs across various classes of algorithms and vendors.
- ☐ Enable faster time to market for multimedia products such as, digital cameras, cell phones, set-top boxes, and portable multimedia players.
- ☐ Provide a standard interface based on given class of multimedia codecs (for example, audio, video, image, and speech).
- ☐ Define common status and parameters based on given class of multimedia codecs.
- ☐ Flexibility of extension of custom functionality.
- ☐ Low overhead of interface.
- ☐ Reduce integration time for system developers.

5.2.6 XDM INTERFACE HISTORY AND ROADMAP

The xDM 0.9 version was released with xDAIS 5.00. xDM 0.9 will continue to be provided and supported for the near term, but is now deprecated. Support for the 0.9 interfaces will be removed in the future. The xDM 1.0 beta version was released with xDAIS 5.10. The major support added in xDM 1.0 is to support for non-blocking process function call.

The xDM 1.0 version is released with xDAIS 5.20. With this 1.0 final release, the 1.0 beta interfaces are no longer supported. For details about differences between xDM versions 0.9 and 1.0 final, see
xDAIS_INSTALL_DIR/packages/ti/xdais/dm/docs/xdm1_differences.pdf.
The xdm1_differences.pdf file contains a list of changes that are likely to be needed to migrate your xDM 0.9-compliant algorithms to xDM 1.0-compliant algorithms.

5.2.7 EXTENDING THE XDM INTERFACES

You can optionally tailor a given algorithm or implementation by extending the xDM interface to create a codec-specific IMOD interface. The algorithm can add more functionality to the xDM interface to define the IMOD interface. The relationship between the xDM and IMOD interfaces are as follows:

- ☐ `IMOD_Fxns`. xDM functions and extension functions

- ☐ `IMOD_Params`. xDM Params (consist of creation and run time) + extension parameters (creation and runtime)
- ☐ `IMOD arguments`. Includes `IMOD_InArgs`, `IMOD_OutArgs`, `IMOD_DynamicParams`, and `IMOD_Status`

Note that the fields in most of these structures changed from xDM version 0.9 to 1.0. See Section 1.3, "xDM Interface History and Roadmap," for more information.

5.2.7.1 Extending an Algorithm

The `ti.xdais.dm.examples.videnc1_copy` example demonstrates how to extend `VIDENC1_InArgs`. The extended structure, which is from `videnc1_copy_ti.h`, is as follows:

```
typedef struct IVIDENC1CPY_InArgs {

IVIDENC1_InArgs videnc1InArgs;

XDAS_Int32 maxBytes; /* Max # of bytes to copy */

} IVIDENC1CPY_InArgs;
```

The implementation of the `process()` function, which uses this optional field, is as follows in `videnc1_copy.c`:

```
/*

* ======== VIDENC1COPY_TI_process ========

*/

XDAS_Int32 VIDENC1COPY_TI_process(IVIDENC1_Handleh,
XDM_BufDesc *inBufs, XDM_BufDesc *outBufs, IVIDENC1_InArgs
*inArgs, IVIDENC1_OutArgs *outArgs)

{

XDAS_Int32 numSamples;
```

```
/* .... */

/* there's an available in and out buffer, how many samples? */

numSamples = inBufs->bufSizes[0] < outBufs->bufSizes[0] ?

inBufs->bufSizes[0] : outBufs->bufSizes[0];

/* honor the extended maxBytes if it was provided */

if (inArgs->size == sizeof(IVIDENC1CPY_InArgs))

   {

   if (numSamples  > ((IVIDENC1CPY_InArgs *)inArgs)->maxBytes)

         {

         numSamples = ((IVIDENC1CPY_InArgs *)inArgs)->maxBytes;

   }

}

/* process the data: read input, produce output */

memcpy(outBufs->bufs[0], inBufs->bufs[0], numSamples);

/* ... */

}
```

5.2.7.2 Extension Considerations for Remotablity

To enable extensions, most xDM structures contain size as their first field. This field is used:

- ☐ By the framework to determine the size of the structure.
- ☐ By the codec to determine how to interpret the fields.

Some frameworks may impose further constraints. For example, the Codec Engine, because it is RPC-based, has the following constraints when using remote codecs:

- No pointers may be used in the extended fields. Because of address translation (the GPP-side address doesn't match the DSPside address) and cache maintenance (the DSP is cached, which requires maintenance for coherence), pointers to data are non-trivial to manage. The default VISA RPC stubs and skeletons manage pointers defined in the base class, but it's impossible for them to know about pointers in proprietary extensions.

CHAPTER 6

Sample Application Using xDM

This chapter was written by Aravindhan K., Member Group Technical Staff, Texas Instruments India Pvt. Ltd.

6.1 OVERVIEW

This chapter shows how to leverage the xDM APIs to create a simple application. We use the H.264 Encoder as an example and detail the sample usage for xDM 1.0 non-blocking support of the process call.

6.2 TEST APPLICATION – OVERVIEW

The test application exercises the IVIDENC base class of the H.264 encoder library.

Figure 6.1 depicts the sequence of APIs exercised in the sample test application.

The test application is divided into four logical blocks:

- Parameter setup
- Algorithm instance creation and initialization
- Process call
- Algorithm instance deletion

6.2.1 PARAMETER SETUP

Each codec component requires various codec configuration parameters to be set at initialization. For example, a video codec requires parameters such as video height, video width, etc. The test application obtains the required parameters from the encoder configuration.

In this logical block, the test application does the following:

- Sets the IVIDENC_Params structure
- Initializes the various DMAN3 parameters
- Reads the input bit stream into the application input buffer
- After successful completion of the above steps, the test application does the algorithm instance creation and initialization.

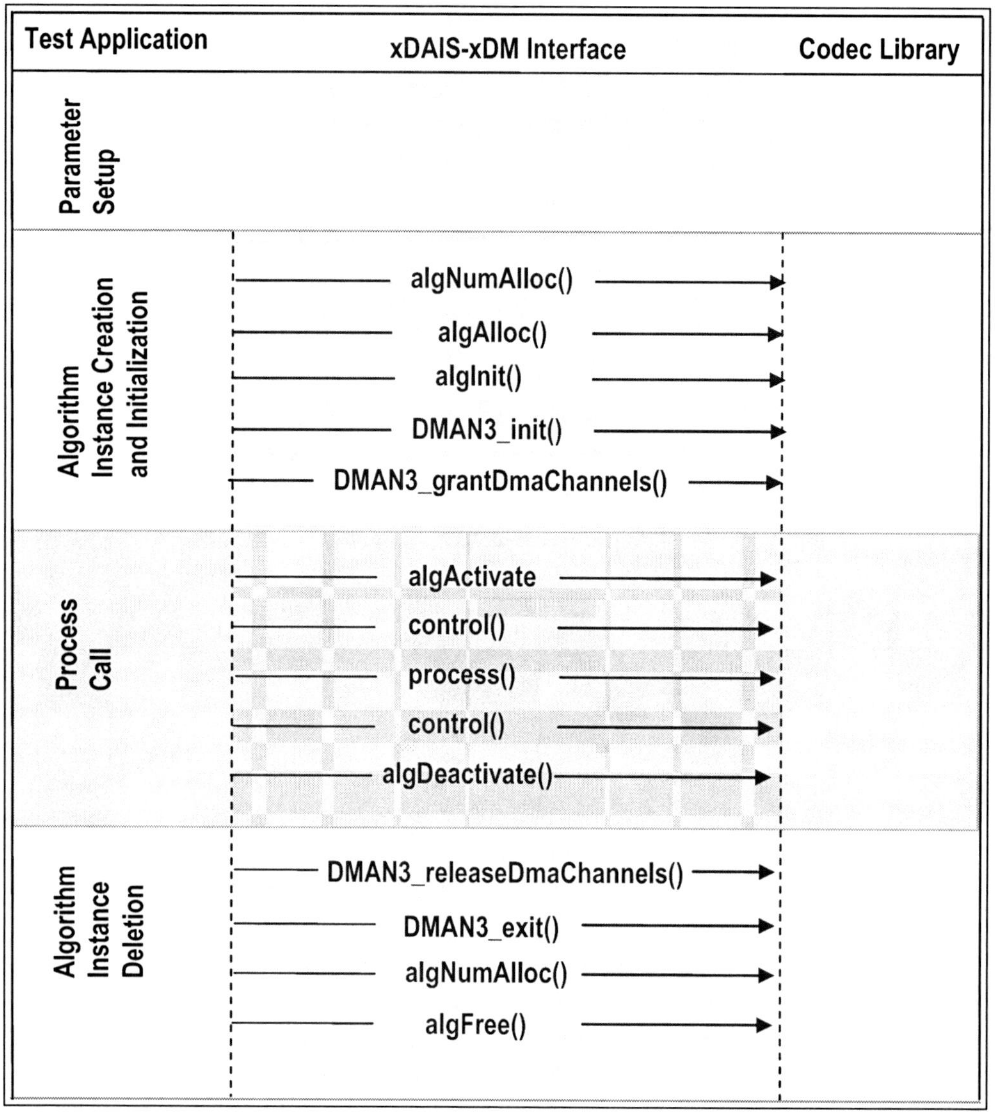

Figure 6.1: Test Application Sample Implementation.

6.2.2 ALGORITHM INSTANCE CREATION AND INITIALIZATION

In this logical block, the test application accepts the various initialization parameters and returns an algorithm instance pointer. The following APIs are called in sequence:

`algNumAlloc()` - To query the algorithm about the number of memory records it requires.

`algAlloc()` - To query the algorithm about the memory requirement to be filled in the memory records.

`algInit()` - To initialize the algorithm with the memory structures provided by the application.

After successful creation of the algorithm instance, the test application does DMA resource allocation for the algorithm. This requires initialization of DMA Manager Module and grant of DMA resources. This is implemented by calling DMAN3 interface functions in the following sequence:

`DMAN3_init()` - To initialize the DMAN module.

`DMAN3_grantDmaChannels()` - To grant the DMA resources to the algorithm instance.

Note:
DMAN3 function implementations are provided in dman3.a64P library.

6.2.3 PROCESS CALL - XDM 0.9

After algorithm instance creation and initialization, the test application does the following:

- Sets the dynamic parameters (if they change during run time) by calling the `control()` function with the `XDM_SETPARAMS` command.

- Sets the input and output buffer descriptors required for the `process()` function call. The input and output buffer descriptors are obtained by calling the `control()` function with the `XDM_GETBUFINFO` command.

- Calls the `process()` function to encode/decode a single frame of data. The behavior of the algorithm can be controlled using various dynamic parameters (see Section Error! Reference source not found.). The inputs to the process function are input and output buffer descriptors, pointer to the `IVIDENC_InArgs` and `IVIDENC_OutArgs` structures.

The `control()` and `process()` functions should be called only within the scope of the `algActivate()` and `algDeactivate()` xDAIS functions which activate and deactivate the algorithm instance, respectively. Once an algorithm is activated, there could be any ordering of `control()` and `process()` functions. The following APIs are called in sequence:

- `algActivate()` - To activate the algorithm instance.

- `control()` (optional) - To query the algorithm on status or setting of dynamic parameters etc., using the six available control commands.

- `process()` - To call the Encoder with appropriate input/output buffer and arguments information.

- `control()` (optional) - To query the algorithm on status or setting of dynamic parameters etc., using the six available control commands.

- `algDeactivate()` - To deactivate the algorithm instance.

The do-while loop encapsulates frame level `process()` call and updates the input buffer pointer every time before the next call. The do-while loop breaks off either when an error condition occurs or when the input buffer exhausts. It also protects the `process()` call from file operations by placing appropriate calls for cache operations as well. The test application does a cache invalidate for the valid input buffers before `process()` and a cache write back invalidate for output buffers after `process()`.

6.2.4 PROCESS CALL - XDM 1.0

xDM 1.0 supports non-blocking implementation for the process function. After algorithm instance creation and initialization, the application does the following:

- Sets the dynamic parameters (if they change during run time) by calling the `control()` function with the `XDM_SETPARAMS` command.

- Sets the input and output buffer descriptors required for the `process()` function call. The input and output buffer descriptors are obtained by calling the `control()` function with the `XDM_GETBUFINFO` command.

- Implements the process call based on the mode of operation – blocking or non-blocking. These different modes of operation are explained below. The behavior of the algorithm can be controlled using various dynamic parameters. The inputs to the `process()` functions are input and output buffer descriptors, pointer to the `IVIDDEC1_InArgs` and `IVIDDEC1_OutArgs` structures.

- Call the `process()` function to encode/decode a single frame of data. After triggering the start of the encode/decode frame start, the video task can be put to SEM-pend state using semaphores. On receipt of the interrupt signal for the end of frame encode/decode, the application should release the semaphore and resume the video task which will do any bookkeeping operations by the codec and updating the output parameters structure -`IVIDDEC1_OutArgs`.

Note:
The process call returns control to the application after the initial setup related tasks are completed.
Application can schedule a different task to use the freed up Host resource.
All service requests from vIMCOP handled via interrupts.
Application resumes the suspended process call after last service request for vIMCOP has been handled.
Application can now complete concluding portions of the process call and return.

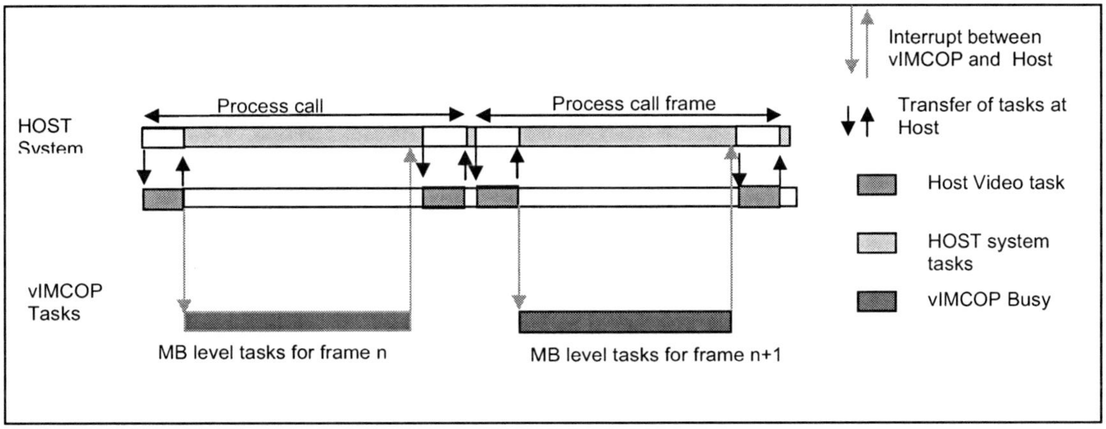

Figure 6.2: Process call with Host release.

The `control()` and `process()` functions should be called only within the scope of the `algActivate()` and `algDeactivate()` xDAIS functions which activate and deactivate the algorithm instance respectively. Once an algorithm is activated, there could be any ordering of `control()` and `process()` functions. The following APIs are called in sequence:

- `algActivate()` - To activate the algorithm instance.
- `control()` (optional) - To query the algorithm on status or setting of dynamic parameters etc., using the six available control commands.
- `process()` - To call the Decoder with appropriate input/output buffer and arguments information.
- `control()` (optional) - To query the algorithm on status or setting of dynamic parameters etc., using the six available control commands.
- `algDeactivate()` - To deactivate the algorithm instance.

The do-while loop encapsulates frame level `process()` call and updates the input buffer pointer every time before the next call. The do-while loop breaks off either when an error condition occurs or when the input buffer exhausts. It also protects the `process()` call from file operations by placing appropriate calls for cache operations as well. The test application does a cache invalidate for the valid input buffers before `process()` and a cache write back invalidate for output buffers after `process()`.

6.2.5 ALGORITHM INSTANCE DELETION

Once encoding/decoding is complete, the test application must release the DMA channels granted by the DMA Manager interface and delete the current algorithm instance. The following APIs are called in sequence:

- `DMAN3_releaseDmaChannels()` - To remove logical channel resources from an algorithm instance.
- `DMAN3_exit()` - To free DMAN3 memory resources.
- `algNumAlloc()` - To query the algorithm about the number of memory records it used.
- `algFree()` - To query the algorithm to get the memory record information.

A sample implementation of the delete function that calls `algNumAlloc()` and `algFree()` in sequence is provided in the `ALG_delete()` function implemented in the `alg_create.c` file.

6.3 FRAME BUFFER MANAGEMENT BY APPLICATION – XDM 1.0

6.3.1 FRAME BUFFER INPUT AND OUTPUT

With the new xDM 1.0, decoder does not ask for frame buffer at the time of `alg_create()`. It uses buffer from "`XDM1_BufDesc *outBufs`" which it gets during each decode process call. Hence, there is no distinction between DPB and display buffers. The framework just needs to ensure that it does not overwrite to buffers, which are locked by codec.

```
H264VDEC_create();

H264VDEC_control(XDM_GETBUFINFO); /* Returns default PAL
                                        D1 size */
do{

H264VDEC_decode(); //call the decode API
```

```
H264VDEC_control(XDM\_GETBUFINFO); /* updates the memory
        required as per the size parsed in stream header */
}

while(all frames)
```

> **Note:**
> App can take that info and change the size of the buffer passed in the next process call.
> It can even re-use the extra buffer space of the 1st frame if the above control call returns a small size that what was given.

The frame pointer given by application and that returned by algorithm may be different. `BufferID(InputID/outputID)` provides the unique ID to keep the record of buffer given to algorithm and released by algorithm. The below figure explains the frame pointer usage.

> **Note:**
> Frame pointer returned by codec in display_bufs will point to actual start location of picture.
> Frame height and width will the actual height and width (after removing cropping and padded width).
> Frame pitch will tell the jump in case to traverse to same pixel location in next line.

As explained above, buffer pointer cannot be used as unique identifier to keep record of frame buffers. Any buffer given to algorithm should be considered locked by algorithm unless the buffer is returned back to application through `IVIDDEC1_OutArgs−>freeBufID[]`.

> **Note:**
> BufferID given back in `IVIDDEC1_OutArgs−>outputID[]` are just for display purpose. Application should not consider it free unless it comes as part of `IVIDDEC1_OutArgs−>freeBufID[]`.

6.3.2 FRAME BUFFER MANAGEMENT BY APPLICATION

The application framework can efficiently manage frame buffers by keeping a pool of free frames from which it gives the decoder empty frames upon request.

The sample application also provides a prototype for managing frame buffers. It implements the following functions, which are defined in file `buffermanager.c` provided along with test application.

`BUFFMGR_Init()`

The `BUFFMGR_Init` function is called by the test application to initialize the global buffer element array to default and to allocate required number of memory data for reference and output buffers. The maximum required dpb size is defined by the supported profile & level.

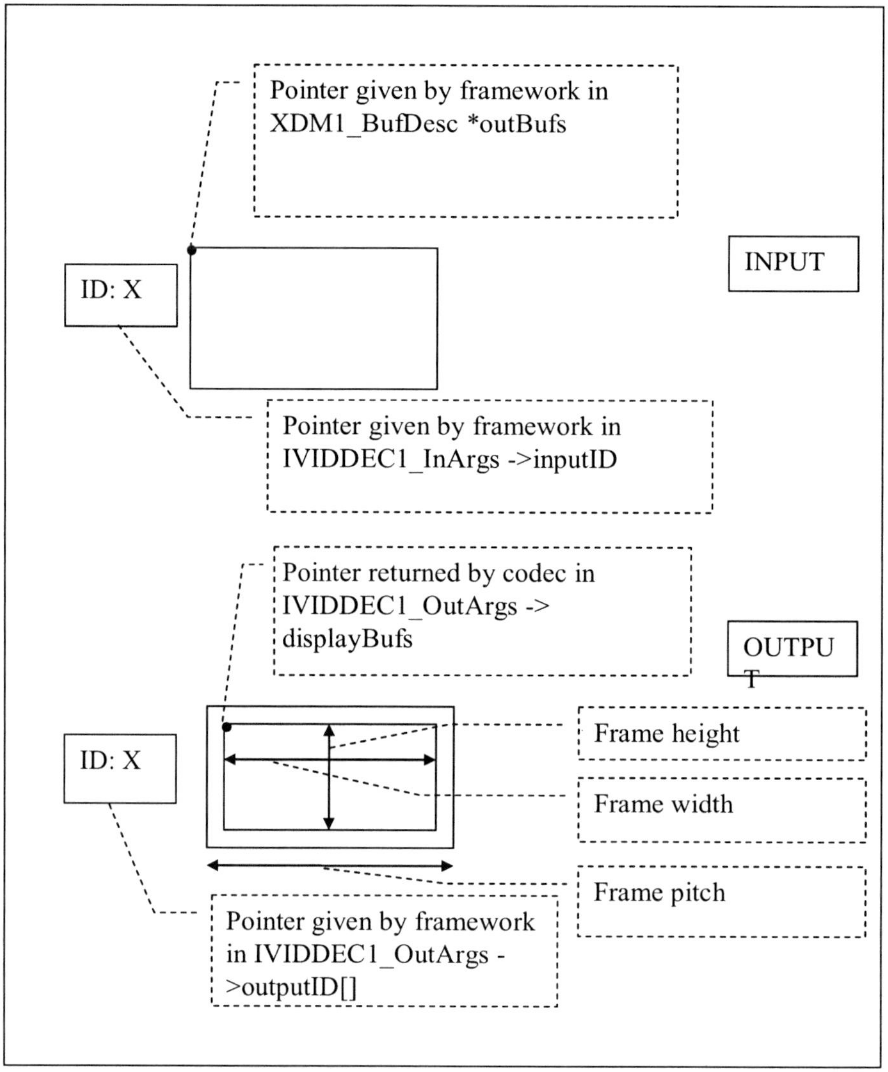

Figure 6.3: Frame buffer pointer Implementation.

`BUFFMGR_ReInit()`

The `BUFFMGR_ReInit` function allocates global luma and chroma buffers and allocates entire space to first element. This element will be used in first frame decode. After the picture's height and width and its luma and chroma buffer requirements are obtained the global luma and chroma buffers are re-initialized to other elements in the buffer array.

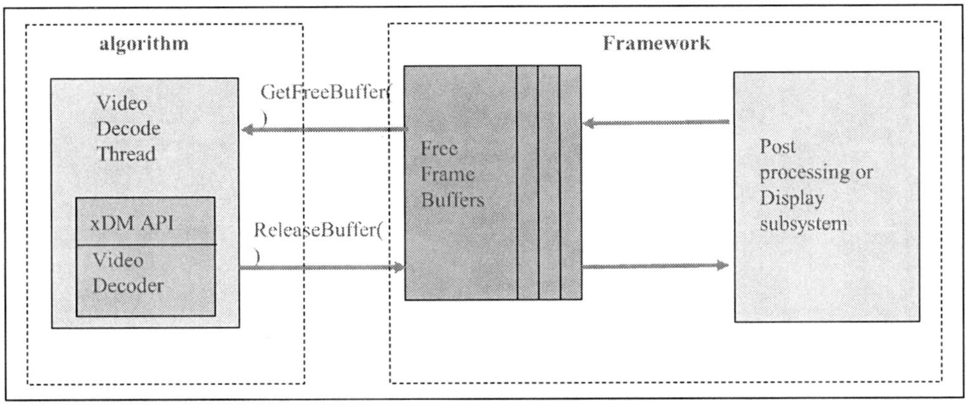

Figure 6.4: Interaction of frame buffers between application and framework.

`BUFFMGR_GetFreeBuffer()`
The `BUFFMGR_GetFreeBuffer` function searches for a free buffer in global buffer array and returns the address of that element. Incase if none of the elements are free then it returns NULL.
`BUFFMGR_ReleaseBuffer()`
The `BUFFMGR_ReleaseBuffer` function takes an array of buffer-ids which are released by the test-app. "0" is not a valid buffer Id hence this function keeps moving until it encounters a buffer Id as zero or it hits the `MAX_BUFF_ELEMENTS`.
`BUFFMGR_DeInit()`
The `BUFFMGR_DeInit` function releases all memory allocated by buffer manager.

6.3.3 HANDSHAKING BETWEEN APPLICATION AND ALGORITHM

Application provides the algorithm with its implementation of functions for video task to go in SEM-pend state when the execution is happening in co-processor. The algorithm calls these application functions to put video task in SEM-pend state.

> **Note:**
> Process call architecture to share Host resource among multiple threads.
> ISR ownership is with the Host layer resource manager – outside the codec.
> The actual codec routine to be executed during ISR is provided by the codec.
> OS/System related calls (SEM_pend, SEM_post) also outside the codec.
> Codec OS agnostic.

The functions to be implemented by applications are:

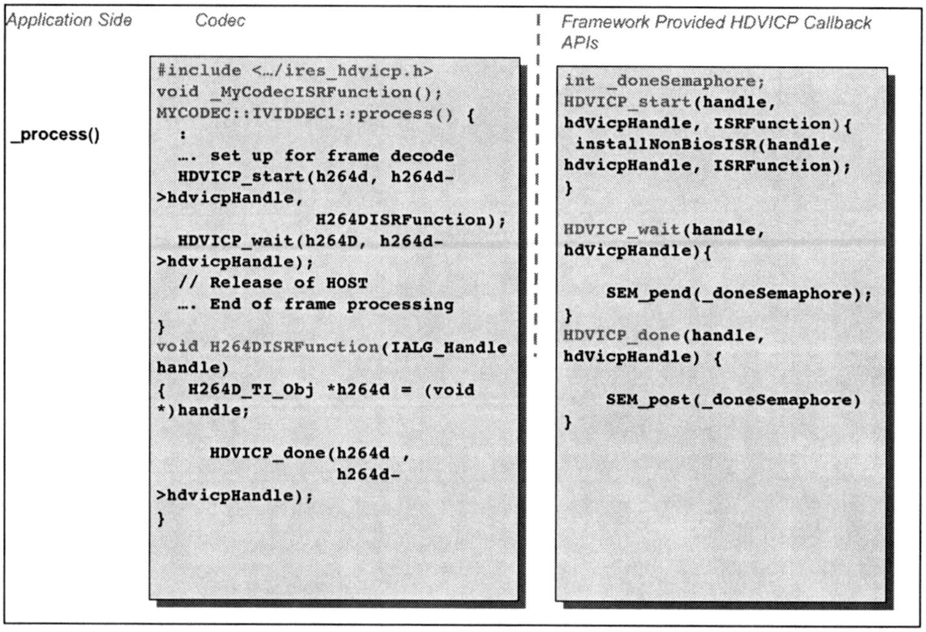

Figure 6.5: Interaction between application and codec.

`HDVICP_initHandle(void *hdvicpHandle)`

This is the top-level function, which initializes hdvicp handle that will be useful when `HDVICP_Wait` and `HDVICP_Done` functions are called by algorithm.

`HDVICP_configure(IALG_Handle handle, void *hdvicpHandle, void (*ISRfunctionptr)(IALG_Handle handle))`

This function is called by algorithm to register its ISR function, which the application needs to call when it receives, interrupts pertaining to video task.

`HDVICP_wait (void *hdvicpHandle)`

This function is called by algorithm to put the video task in SEM-pend state.

`HDVICP_done (void *hdvicpHandle)`

This function is called by algorithm to release the video task from SEM-pend state.

In the sample test application `HDVICP_wait()` is implemented using polling. The application can implement it in a way considering the underlying system.

Interrupts from ARM968 to Host ARM926 is used to inform when the frame processing is done. vIMCOP sends interrupt which maps to INT No 28 (KALINT9 Video IMCOP) of ARM926

INTC. Other interrupts INT NO 19-27(KALINT0 – KALINT8) should be disabled in ARM926 INTC.

Sample test application has set a priority of "1" for this interrupt and hence uses FIQ to service it. In the actual application, the system integrator can choose the priority depending on its need.

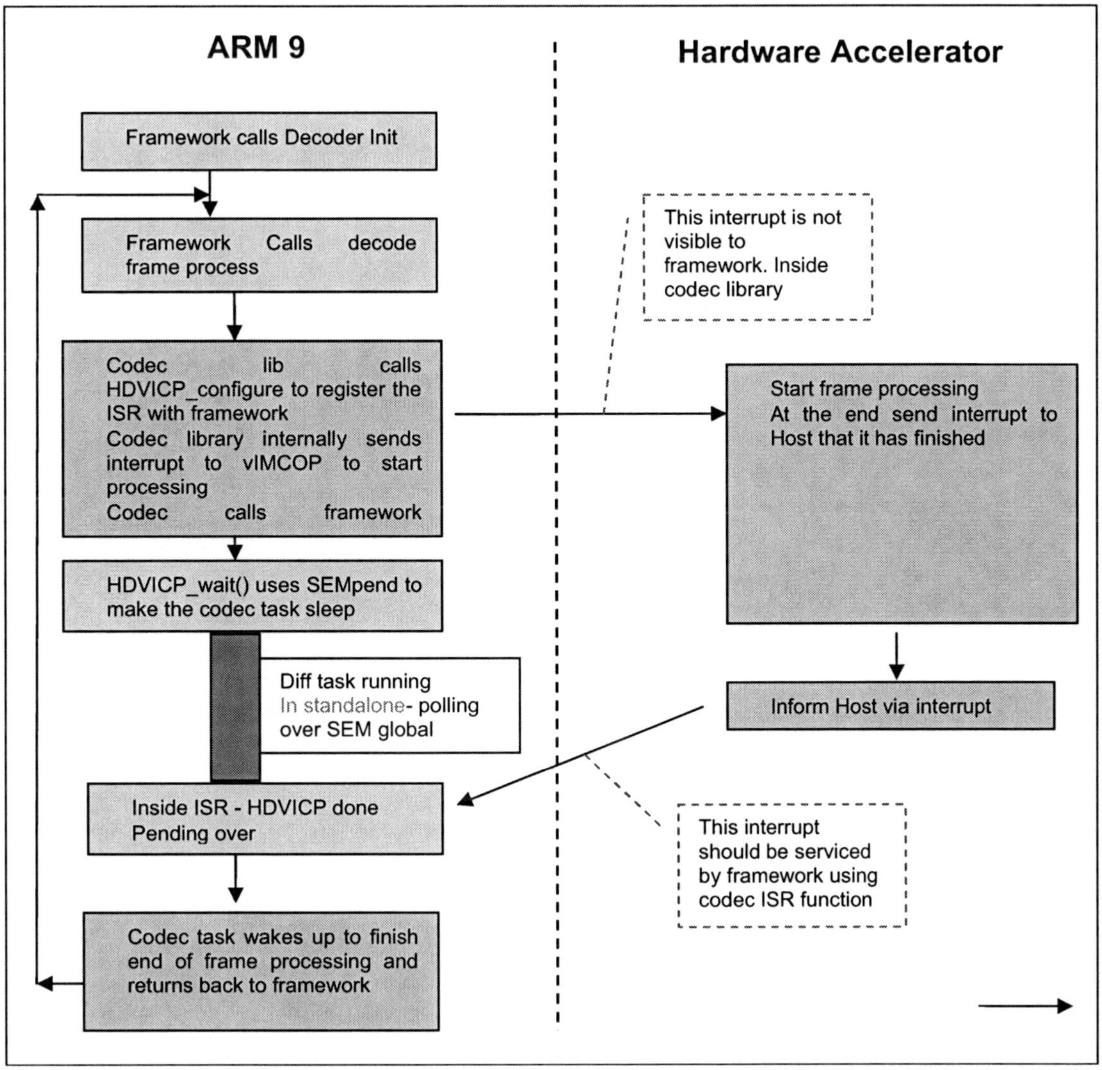

Figure 6.6: Interrupt between codec and application.

6.3.4 SAMPLE TEST APPLICATION

The test application exercises the IVIDDEC1 base class of the H.264 Decoder.

```
/* Main Function acting as a client for Video Decode Call    */

ARM926_enable_FIQ();  /* SWI call to enable interrupts      */

ARM926_INTC_init();  /* Init AINTC */

BUFFMGR_Init();

TestApp_SetInitParams(&params.viddecParams);

/* Init the DMA param */

TestApp_SetDMAInitParams();

/* Init Hdvicp params */

HDVICP_initHandle(&hdvicpObj);

/*------------------- Decoder creation ----------------------*/

handle = (IALG_Handle) H264VDEC_create();

/* Get Buffer information              */

H264VDEC_control(handle, XDM_GETBUFINFO);

/* Do-While Loop for Decode Call  for a given stream  */

  do

  {

/* Read the bitstream in the Application Input Buffer */

 validBytes = ReadByteStream(inFile);
```

```
/* Get free buffer from buffer pool */

   buffEle = BUFFMGR_GetFreeBuffer();

/* Optional: Set Run time parameters in the Algorithm via control() */

  H264VDEC_control(handle, XDM_SETPARAMS);

/*-------------------------------------------------------*/

/* Start the process : To start decoding a frame        */

/* This will always follow a H264VDEC_decode_end call */

/*-------------------------------------------------------*/

  retVal = H264VDEC_decode

  (

       handle,

       (XDM1_BufDesc *)&inputBufDesc,

       (XDM_BufDesc *)&outputBufDesc,

       (IVIDDEC1_InArgs *)&inArgs,

       (IVIDDEC1_OutArgs *)&outArgs

   );

/* Get the statatus of the decoder using comtrol */

H264VDEC_control(handle, IH264VDEC_GETSTATUS);

/* Get Buffer information :         */

H264VDEC_control(handle, XDM_GETBUFINFO);
```

```
/* Optional: Reinit the buffer manager in case the

/* frame size is different              */

BUFFMGR_ReInit();

/* Always release buffers - which are released from

/* the algorithm side -back to the buffer manager   */

BUFFMGR_ReleaseBuffer((XDAS_UInt32 *)outArgs.freeBufID);

} while(1);

/* end of Do-While loop - which decodes frames            */

ALG_delete (handle);

BUFFMGR_DeInit();
```

Note:
This sample test application does not depict the actual function parameter or control code. It just shows the basic flow of the code.

In compliance to xDM 1.0 - application now needs to supply buffers to codec module for filling in the reference frames and output buffers for display. For this purpose, the application defines and invokes a set of functions for buffer management. In addition to this, the application also invokes the standard functions, which are exposed by the algorithm through the xDM1.0 API. In every process call to the codec module application layer provides one free buffer identified by a unique buffer Id. ***Please note that 0 is an invalid buffer Id. Buffer ids are positive non-zero numbers.***

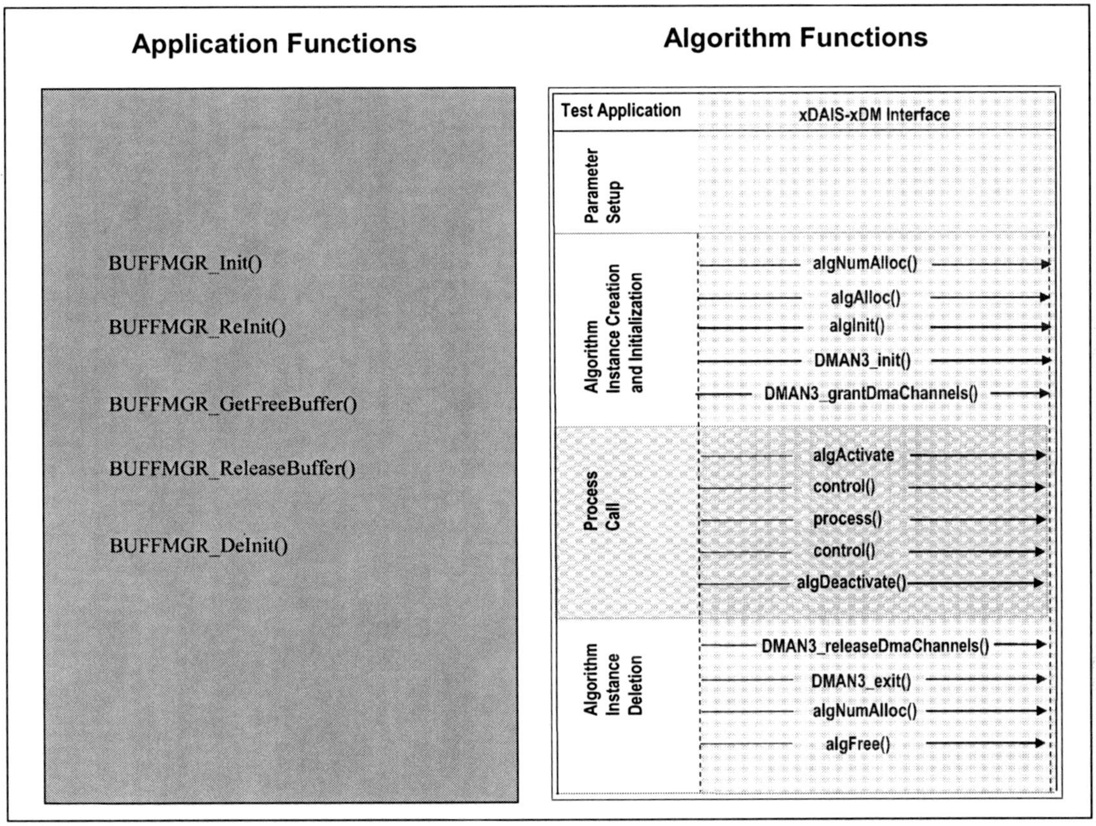

Figure 6.7: Buffer manager and its interaction with xDM interface.

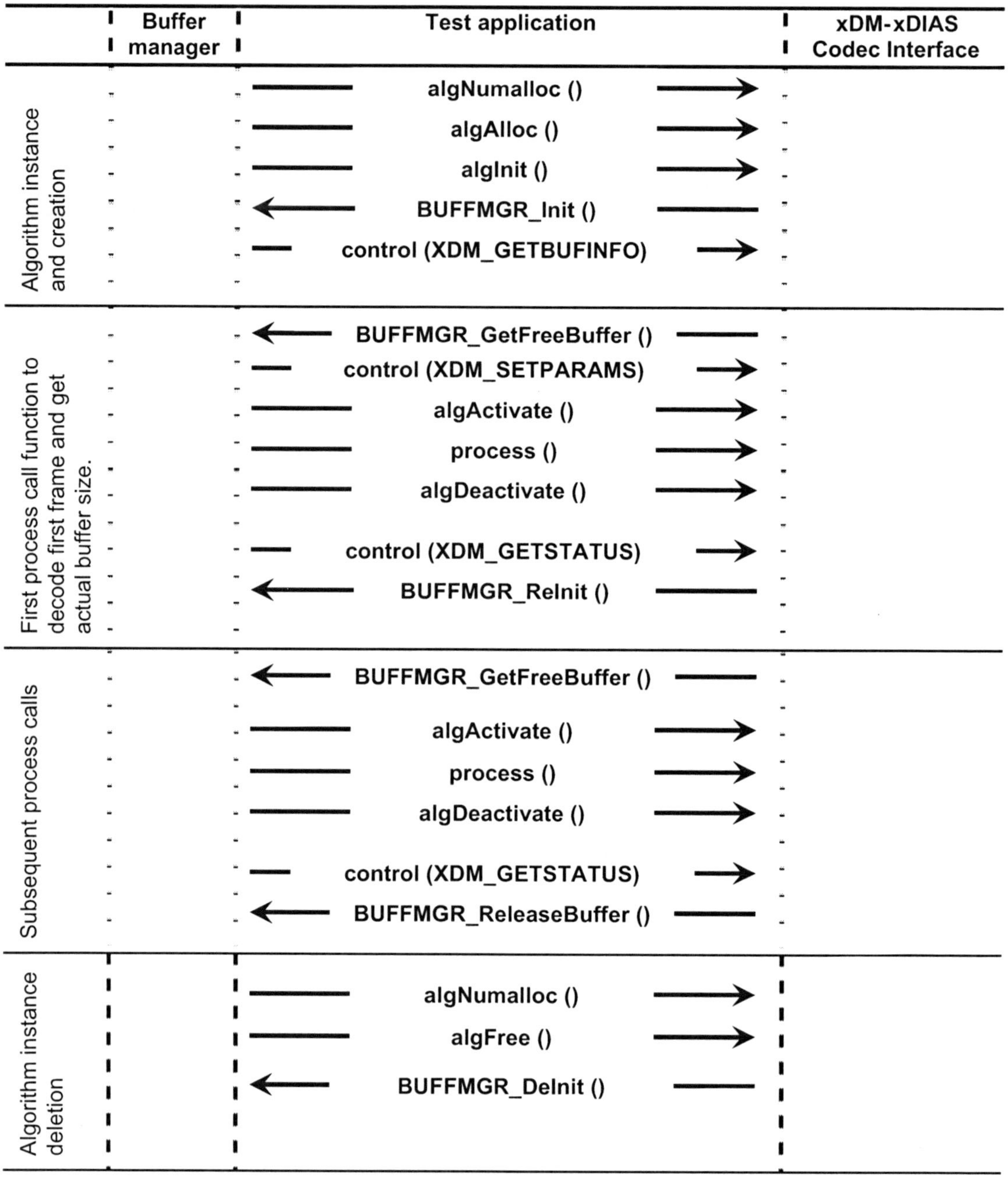

Figure 6.8: Function Call Sequence.

CHAPTER 7

Embedded Peripheral Software Interface (EPSI)

This chapter was written with Anand Balagopalakrishnan, Texas Intruments India Pvt. Ltd.

7.1 OVERVIEW

The DaVinci software framework provides standard Linux driver interfaces that control and configure the peripherals attached to the Silicon. These APIs abstract the application from the hardware details of the peripherals. An application written using these APIs can be easily ported to another Linux based DaVinci Silicon like DM355 or DM6467. However, these APIs are specific to Linux. Therefore, when the application is ported to a DSP/BIOS based DaVinci product like DM6437 or DM648, quite a few changes will be required in the application.

This chapter extends the concept of standard Linux APIs and explains how the user can define an additional layer on top of the OS specific driver APIs. This layer makes the driver APIs practically OS agnostic. As an example, we define a video capture interface that is applicable to both Linux and DSP/BIOS.

Note:

This chapter shows "a" method of extending device driver APIs to get a standard Peripheral Interface across platforms.

Reference code provided is to facilitate better understanding of the concept. It may or may not have actual correspondence with the DVSDK.

Error checking is intentionally left out of the reference code to improve readability.

7.2 INPUT / OUTPUT LAYER

The Input Output Layer in the DaVinci software framework provides services for configuring and controlling the peripherals attached to the device.

The driver APIs on DaVinci Linux consists of:

- ☐ Standard Linux driver APIs such as V4L2, FBDev for video and OSS for audio
- ☐ TI proprietary driver interface for McBSP, CMEM, DSP Link

Linux application developers need to learn only the Linux driver APIs to control a peripheral. Application programmers need not be concerned with the details involved in implementing a specific

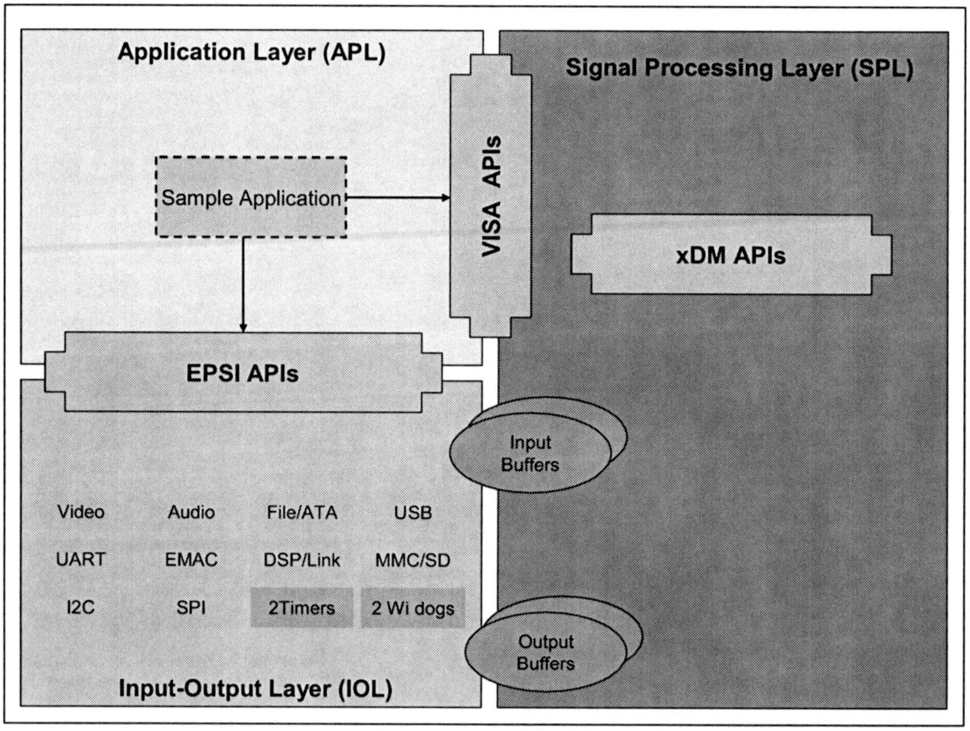

Figure 7.1: DaVinci Software Framework.

peripheral. An application developed using the driver APIs will remain unchanged if a peripheral IP block is replaced with another. The configuration parameters might change for the new peripheral block but the basic interface will remain the same. Thus, an application written using the standard Linux APIs (say V4L2 for video) remains more or less unchanged when the application is ported from one Linux platform to another—for example: DM6446 – > DM355 – > DM6467.

As seen above, these services are provided by device drivers. A device driver is a special piece of system software written for a device to initialize, configure and perform IO operations on that device. These device drivers are dependent on the Operating Systems. Each OS (Linux, DSP/BIOS, Win CE, etc.) has a specific interface that a device driver developer needs to adhere to. The "look and feel" of the device driver APIs vary between OS. For example, the device driver APIs for Linux are different from the device driver APIs for DSP/BIOS.

Usage of device driver APIs creates portability issues when an application is migrated from one OS to another – say DM6446 with Linux to DM6437 with DSP/BIOS. However, in spite of the differences in device drivers between OS, the policy adopted by device drivers remains similar.

It would be possible to define a common interface (EPSI) across all OS and have a separate glue layer that maps the EPSI APIs to device driver specific APIs.

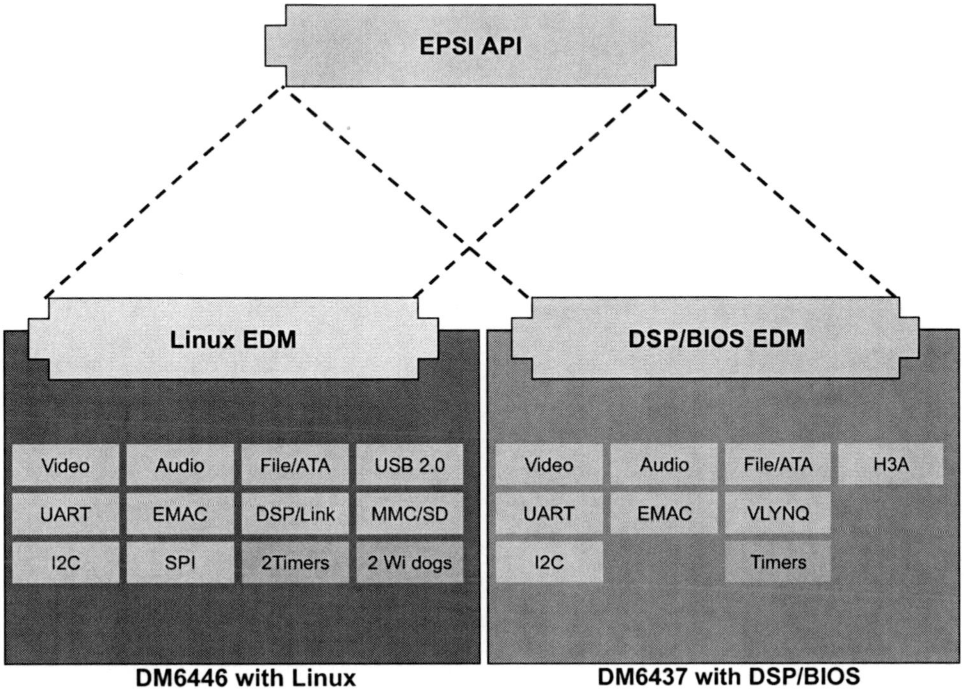

Figure 7.2: EPSI and EDM.

The EPSI interface is common across both Linux on DM6446 and DSP/BIOS on DM6437. Each OS has a separate glue layer called EPSI to Driver Mapping (EDM) for each device. The Linux EDM glue layer maps the EPSI APIs to Linux specific device driver APIs, and similarly the DSP/BIOS EDM glue layer does the same for DSP/BIOS APIs. Also, note that the devices available on DM6446 and DM6437 are different.

Definition of EPSI APIs does not mask or prevent the usage of device driver APIs directly. The choice of using the EPSI APIs or the device driver APIs lies with the application developer:

- Application programmers who need portability can use EPSI APIs in their application. However, the EDM glue layer adds a small performance overhead. In addition, the device driver APIs may provide more device specific configurations than comprehended by the EPSI.

- Application programmers who need minimum latency and maximum configurability can use the device driver APIs. In this case, the application developers lose on portability when migrating the application to another OS.

This chapter specifies the EPSI APIs suggested for use by application developers. It also shows how the EDM glue layer can be written for the VPFE (Video Peripheral Front End) used for video capture.

7.3 EPSI APIS

This section lists the EPSI APIs suggested for any device. For a generic device DEV, the following EPSI APIs abstract the device driver APIs (see Table 7.2).

Table 7.2: Generic EPSI APIs

API	Description
DEV_open()	Initializes the device and returns a handle
DEV_read()	Reads from the device to the buffers provided by application
DEV_write()	Writes to the device from buffers provided by application
DEV_close()	Uninitializes the device and closes the handle
DEV_control()*	Configures the device
DEV_getBuffer()#	Gets the next ready buffer from the device queue for application consumption
DEV_returnBuffer()#	Returns the buffer to the device queue after application consumption

*—Actual configuration parameters and values depend on the device and platform

#—Optional APIs used when the device driver uses queuing model to manage buffers

7.4 EDM FOR LINUX

Devices or peripherals are treated as special files in Linux. The devices are operated upon using the standard Linux file descriptors (FD). All the standard file operations such as open(), read(), write(), ioctl() and close() are also applicable for Linux device driver APIs.

Each EPSI API has a corresponding EDM function defined. These function definitions are given in the subsections below. Error checking and initializations are removed from the function definitions for readability.

7.4.1 VPFE_OPEN

This function sets up the VPFE peripheral and returns a handle to the VPFE module. This handle is used in subsequent calls to VPFE module. To setup a VPFE peripheral, this function uses V4L2 APIs on Linux.

This function opens the VPFE device (typically /dev/v4l/video0) and obtains a file descriptor for the VPFE. The FD is then used to configure the video input (composite, component, s-video), video standard (NTSC, PAL, SECAM, AUTO), video file format (UYVY, YUYV, YUV420, etc.). These configurations represent the actual physical connection and the file format supported by the driver. The Linux VPFE driver on DaVinci supports only UYVY or YUYV format.

After configuring the device, the function requests four buffers from the VPFE driver. These buffers are memory mapped to given application buffers in user space. Now, the application can directly use these buffers to process the video frames captured at the front end.

```
VpfeHandle VPFE_open(VPFE_Params *vpfeParams)
{

...

/* Open a handle to FVID device */

vpfeHdl->fd = open(vpfeParams->device, O_RDWR | O_NONBLOCK, 0);

/* Set the video input - Composite/S-Video/Component */

ioctl(vpfeHdl->fd, VIDIOC_S_INPUT, &vpfeParams->videoInput);

/* Set the video standard - NTSC/PAL/SECAM/AUTO */

ioctl(vpfeHdl->fd, VIDIOC_S_STD, &vpfeParams->videoStd);

/* Set the video format - UYVY/YUYV/YUV420/... */

ioctl(vpfeHdl->fd, VIDIOC_S_FMT, &v4l2Format);

/* Request for 4 memory mapped buffers from V4L2 */

v4l2Request.count = 4;

v4l2Request.type = V4L2_BUF_TYPE_VIDEO_CAPTURE;

v4l2Request.memory = V4L2_MEMORY_MMAP;
```

```
ioctl(vpfeHdl->fd, VIDIOC_REQBUFS, &v4l2Request);

/* Memory map the VPFE buffers to application buffers */

for (numBufs=0; numBufs < v4l2Request.count; numBufs++)

{

    v4l2Buf.type = V4L2_BUF_TYPE_VIDEO_CAPTURE;

    v4l2Buf.memory = V4L2_MEMORY_MMAP;

    v4l2Buf.index = numBufs;

    ioctl(vpfe->capFd, VIDIOC_QUERYBUF, &v4l2Buf);

    vpfeHdl->buffers[numBufs].length = v4l2Buf.length;

    vpfeHdl->buffers[numBufs].start = mmap(NULL,
      v4l2Buf.length, PROT_READ | PROT_WRITE,
      MAP_SHARED, vpfeHdl->fd, v4l2Buf.m.offset);

  }

/* Start the video streaming */

v4l2BufType = V4L2_BUF_TYPE_VIDEO_CAPTURE;

ioctl(vpfeHdl->fd, VIDIOC_STREAMON, &v4l2BufType);

return vpfeHdl;

}
```

7.4.2 VPFE_GETBUFFER

The VPFE driver has a queue of buffers for capturing video frames. Whenever a buffer is filled up, it is moved to the back of the queue. This function checks if a buffer is available at the back of the queue – in other words, it checks if there is a buffer that is filled up recently. This is done through the select() system call in Linux. When the select() system call unblocks, it means that there is a buffer ready for use The function then dequeues this buffer from the buffer queue using the ioctl() system call.

Once the buffer is dequeued, it is available to the application for further processing. The dequeued buffer is accessed via vpfeHdl− >buf structure.

```
VpfeStatus VPFE_getBuffer(VPFE_Handle vpfeHdl, char *buff)

{

...

/* Poll the FDs to check if a buffer is available */

select(vpfeHdl->fd + 1, &fds, NULL, NULL, &tv);

/* Remove the buffer from the VPFE queue */

ioctl(vpfeHdl->fd, VIDIOC_DQBUF, &vpfeHdl->buf);

buff = vpfeHdl->buffers[vpfeHdl->buf.index].start;

...

}
```

7.4.3 VPFE_RETURNBUFFER

This function is called by the application to return the buffer to the VPFE buffer queue. The application gets a buffer from the VPFE queue by calling VPFE_getBuffer(). This buffer is then used by the application for further processing—for example, the buffer from VPFE is passed as input to the video encoder.

After the application has finished processing the buffer, it needs to return the buffer back to the VPFE queue. The application does this by calling the VPFE_returnBuffer().

```
VpfeStatus VPFE_returnBuffer(VPFE_Handle vpfeHdl, char *buff)

{
```

```
...

/* Add the buffer to VPFE queue */

ioctl(vpfeHdl->fd, VIDIOC_QBUF, &vpfeHdl->buf);

...

}
```

7.4.4 VPFE_CLOSE

This function is used to uninitialize the VPFE device. Streaming is turned off for this device using ioctl() system call.

The device is then uninitialized using the standard close() system call. In addition, the buffers mapped from the VPFE driver space to the user space are unmapped using the munmap() system call.

```
VpfeStatus VPFE_close(VPFE_Handle vpfeHdl)

{

...

/* Turn off streaming */

v4l2BufType = V4L2_BUF_TYPE_VIDEO_CAPTURE;

ioctl(vpfeHdl->fd, VIDIOC_STREAMOFF, &v4l2BufType);

/* Close the VPFE device */

close(vpfeHdl->fd);
```

```
/* Unmap the memory mapped buffers */

for (i=0; i<4; i++)

{

    munmap(vpfeHdl->buffers[i].start, vpfe->buffers[i].length);

}

...

}
```

7.5 EDM FOR DSP/BIOS

As we saw in Section 3, Linux requires that devices are handled like special files, and all standard APIs for file operations are provided by device drivers also. On the other hand, DSP/BIOS requires that the device drivers are implemented as SIO or GIO module.

We had already looked at the EDM layer for Linux. This section will describe how the EDM layer is defined for DSP/BIOS.

The VPFE device driver for DSP/BIOS is written as a GIO module. The DSP/BIOS PSP provides a wrapper over the GIO functions called as FVID so as to provide the user an easy interface.

7.5.1 VPFE_OPEN

This function sets up the VPFE peripheral and returns a handle to the VPFE module. This handle is used in subsequent calls to VPFE module.

This function opens the VPFE device configured in the DSP/BIOS Tconf script and obtains a handle for the VPFE. The FVID handle is then used to configure the video input (composite, component, s-video), video standard (NTSC, PAL, SECAM, AUTO), video file format (UYVY, YUYV, YUV420, etc.). These configurations represent the actual physical connection and the file format supported by the driver. The DSP/BIOS VPFE driver on 64LC supports only UYVY or YUYV format.

After configuring the device, the function requests for allocation of 4 buffers from the VPFE driver. These buffers are queued at the VPFE driver.

```
VpfeHandle VPFE_open(VPFE_Params *vpfeParams)

{

...

/* Open a handle to VPFE device */

vpfeHdl->fvidHdl = FVID_create("/VPFE0", IOM_INOUT, NULL,
&channelParams, NULL);

/* Configure the VPFE parameters */

/* Set the video input - Composite/S-Video/Component */

/* Set the video standard - NTSC/PAL/SECAM/AUTO */

/* Set the video format - UYVY/YUYV/YUV420/... */

FVID_control(   vpfeHdl->fvidHdl,
VPFE_ExtVD_BASE+PSP_VPSS_EXT_VIDEO_DECODER_CONFIG,
&vpfeParams->tvp5146Params));

/* Request memory for 4 buffers from VPFE */

for (numBufs=0; numBufs < 4; numBufs++)

{
   FVID_alloc(vpfeHdl->fvidHdl, vpfeHdl->buffers[i]);

   FVID_queue(vpfeHdl->fvidHdl, vpfeHdl->buffers[i]);
}

FVID_dequeue(vpfeHdl->fvidHdl, &vpfeHdl->buf);

return vpfeHdl;

}
```

7.5.2 VPFE_GETBUFFER

The VPFE driver has a queue of buffers for capturing video frames. Whenever a buffer is filled up, it is moved to the back of the queue.

The application can obtain a buffer using FVID_dequeue() API. If a buffer is dequeued from the VPFE queue, the buffer depth at the VPFE driver decreases. Hence, it is usual for the application to queue one of its free buffers to the VPFE buffer queue through FVID_queue() call. Instead of this FVID_dequeue() and FVID_queue() pair of API calls, the FVID APIs provide another API – FVID_exchange(). This API removes a buffer from the VPFE buffer queue, takes a buffer from application and adds it buffer to the VPFE buffer queue. The buffer dequeued from the VPFE buffer queue is returned to the application.

This function exchanges a processed buffer for a buffer that is ready for processing. Once the buffer is exchanged, it is available the application for further processing. The buffer to be processed is accessed via vpfeHdl– >buf structure.

```
VpfeStatus VPFE_getBuffer(VPFE_Handle vpfeHdl, char *buff)

{

...

FVID_dequeue(vpfeHdl->fvidHdl, &vpfeHdl->buf);

buff = vpfeHdl->buf;

...

}
```

7.5.3 VPFE_RETURNBUFFER

In Linux, the VPFE_getBuffer() dequeues a buffer from the VPFE queue. It needs to be returned back to the VPFE queue using VPFE_returnBuffer() function.

However, in DSP/BIOS, the buffer is dequeued and queued at once using the FVID_exchange() API inside the VPFE_getBuffer() IOL API. Hence, there is no need to return the buffer back to the VPFE queue.

```
VpfeStatus VPFE_returnBuffer(VPFE_Handle vpfeHdl, char *buff)
{
...

return VPFE_SUCCESS;

...
}
```

7.5.4 VPFE_CLOSE

This function is used to uninitialize the VPFE device. The buffers allocated at the VPFE driver are freed using the FVID_free() API. The device is then uninitialized using FVID_delete() API.

```
VpfeStatus VPFE_close(VPFE_Handle vpfeHdl)

{

...

/* Free the buffers */

for (i=0; i<4; i++)

{
   FVID_free(vpfeHdl->fvidHdl, vpfeHdl->buffers[i]);
}

/* Close the VPFE device */

FVID_delete(vpfeHdl->fvidHdl);

...

}
```

7.6 SUMMARY

The figures below summarize the EPSI APIs for Video Capture (VPFE) and shows how it maps to Linux EDM and DSP/BIOS EDM.

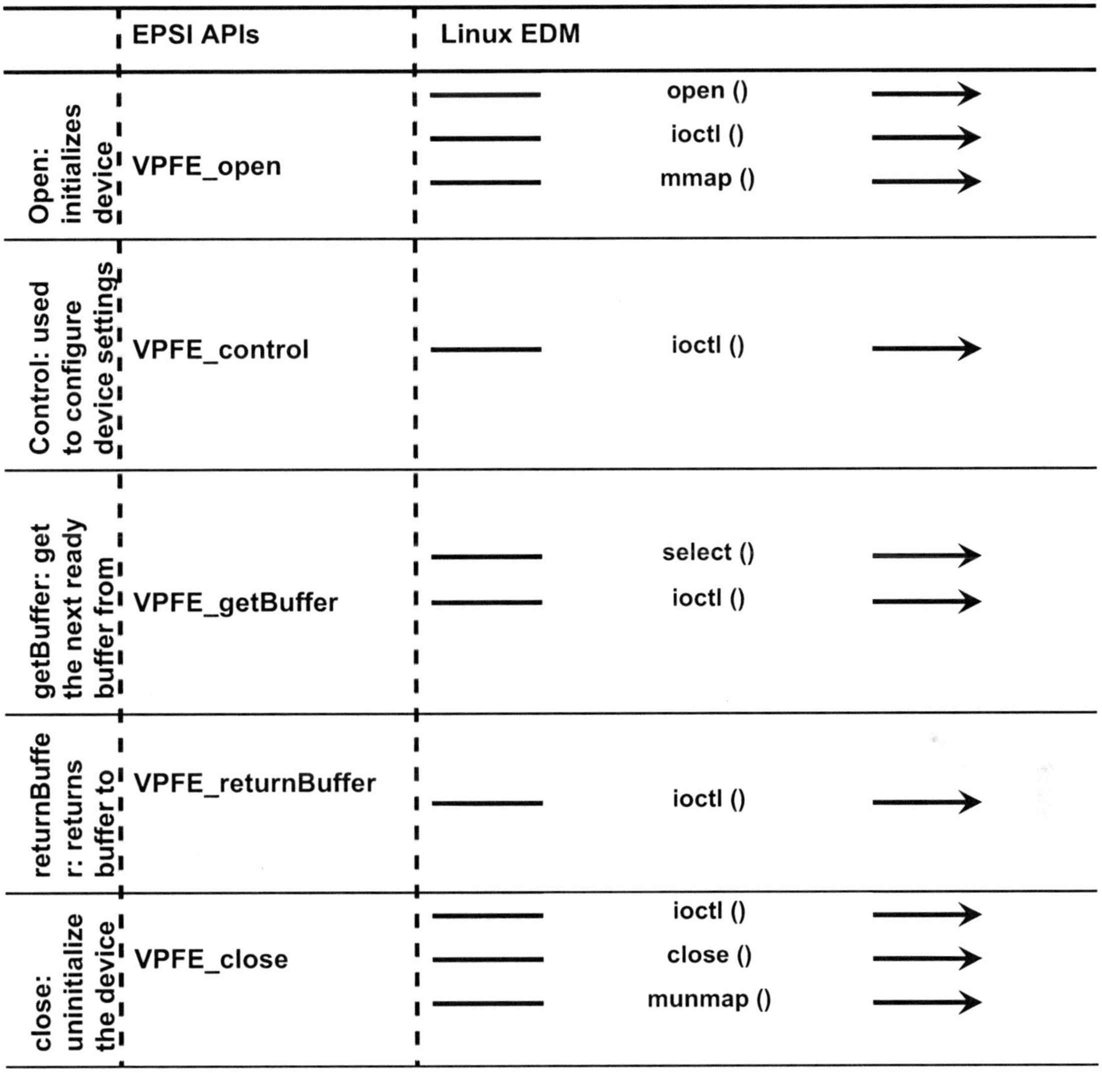

Figure 7.3: EPSI and Linux EDM.

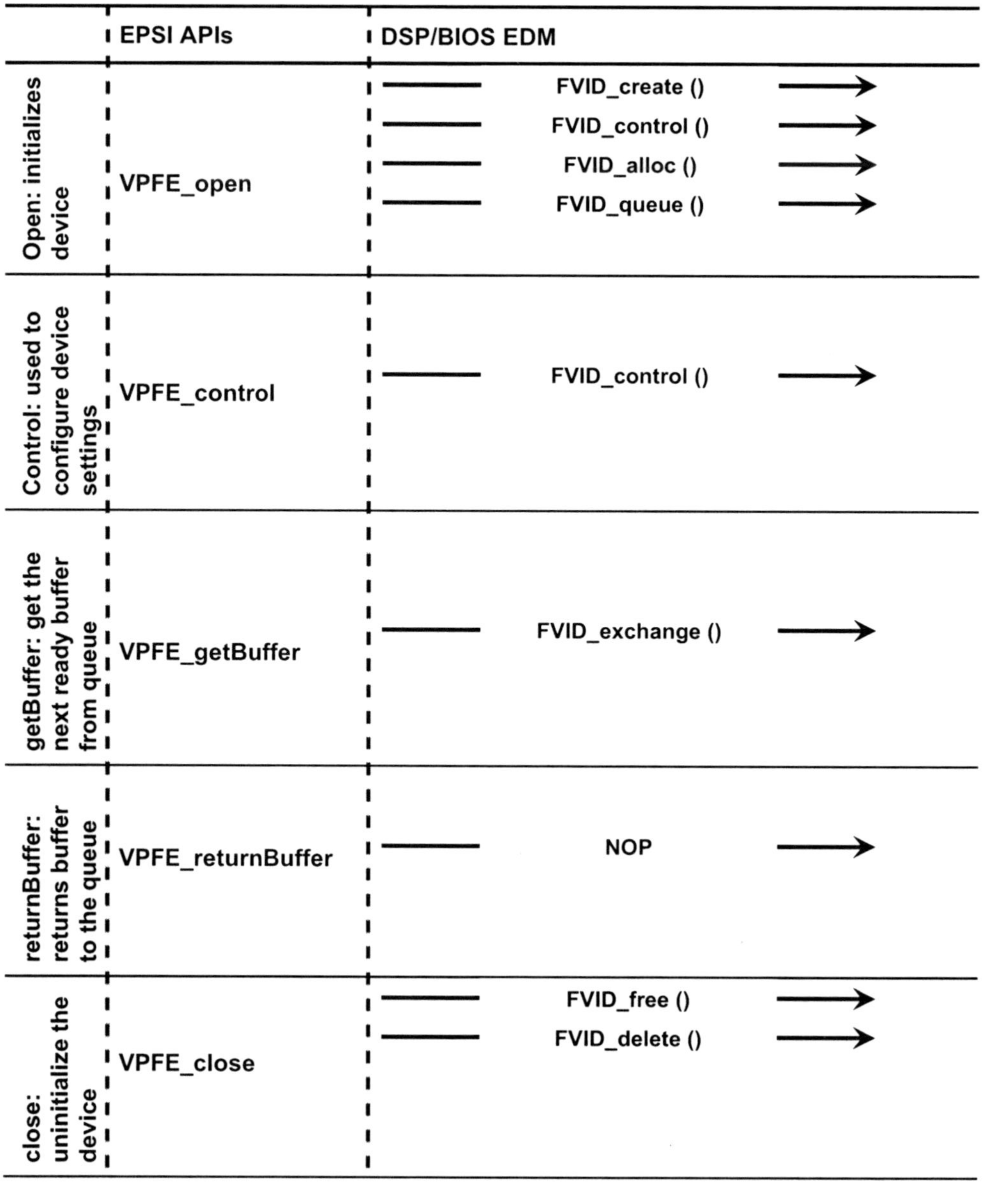

Figure 7.4: EPSI and DSP/BIOS EDM.

CHAPTER 8

Sample Application Using EPSI

This chapter was written by Anand Balagopalakrishnan, Texas Intruments India Pvt. Ltd.

8.1 OVERVIEW

This chapter shows how to use the EPSI APIs to create a simple video capture application. This application captures raw video frames from the video input device using the EPSI APIs explained before.

8.2 VIDEO CAPTURE APPLICATION

The Video Capture application uses the EPSI APIs for VPFE to capture video. The components used by this application are shown in Figure 8.1.

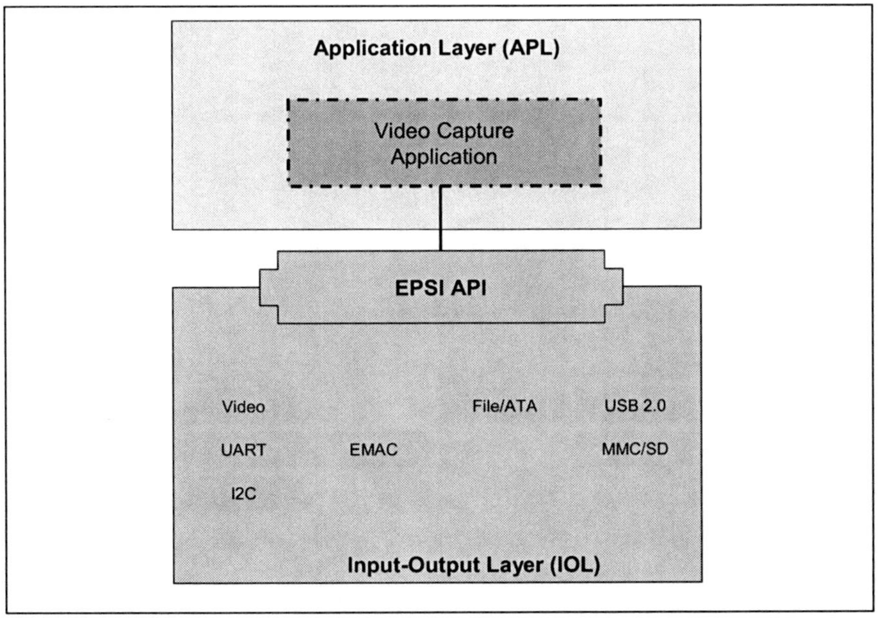

Figure 8.1: EPSI Application.

The Video Capture application captures raw video frames from the video input peripheral attached to the SoC. The application uses only the EPSI APIs for the video capture. The following device drivers are used for this operation:

- Video- VPFE driver
- ATA- Storage driver
- USB- USB driver to capture video to USB based media
- UART- Serial port driver for console debugging, if required
- EMAC-Ethernet driver in case the captured video is sent over the network
- MMC- MMC/SD driver to capture video to MMC/SD media
- I2C - I2C driver used internally by the VPFE driver

Though these drivers are actually in operation, the application need not know about these driver APIs. The application is only required to use two sets of EPSI APIs – one for VPFE and the other for FILE I/O.

In addition, the application uses the same EPSI APIs regardless of the OS on which it is running.

8.3 APPLICATION CONTROL FLOW

The application captures raw video frames from the video input peripheral. The frames captured are stored in a file.

The control flow of the video capture application is shown in Figure 8.2.

The following is the sequence of operations:

1. Initialize video capture device using VPFE_open(). The device is configured with a set of initialization parameters provided by the user.

2. Configure the video capture device using VPFE_control(). This is an optional step. The video device would have been initialized in Step 1. Any further configuration required is handled by VPFE_control().

3. Capture a video frame using VPFE_getBuffer(). This function dequeues a buffer from the VPFE buffer queue which contains the video frame captured.

4. Write the frame into a file using FILE_write(). The actual write time will depend on the data transfer rate of the target media.

5. Return the buffer back to the VPFE buffer queue using VPFE_returnBuffer().

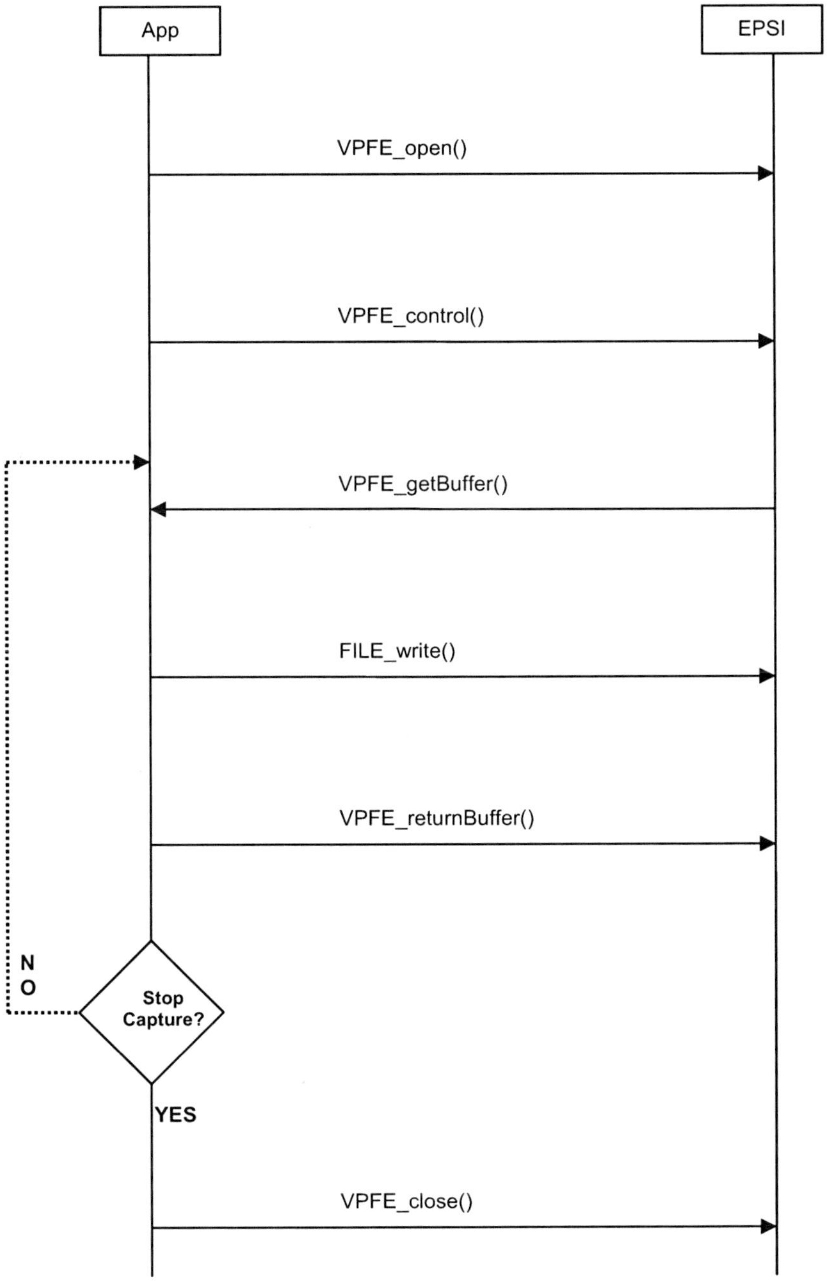
App
EPSI
VPFE_open()
VPFE_control()
VPFE_getBuffer()
FILE_write()
VPFE_returnBuffer()
N
O
Stop
Capture?
YES
VPFE_close()

Figure 8.2:

6. Check if more frames need to be captured. If yes, go to Step 3.

7. Close the VPFE device using VPFE_close().

8.4 APPLICATION IMPLEMENTATION

The implementation of the application follows exactly the control flow shown in previous section. The pseudo code for the application is shown in Figure 8.3.

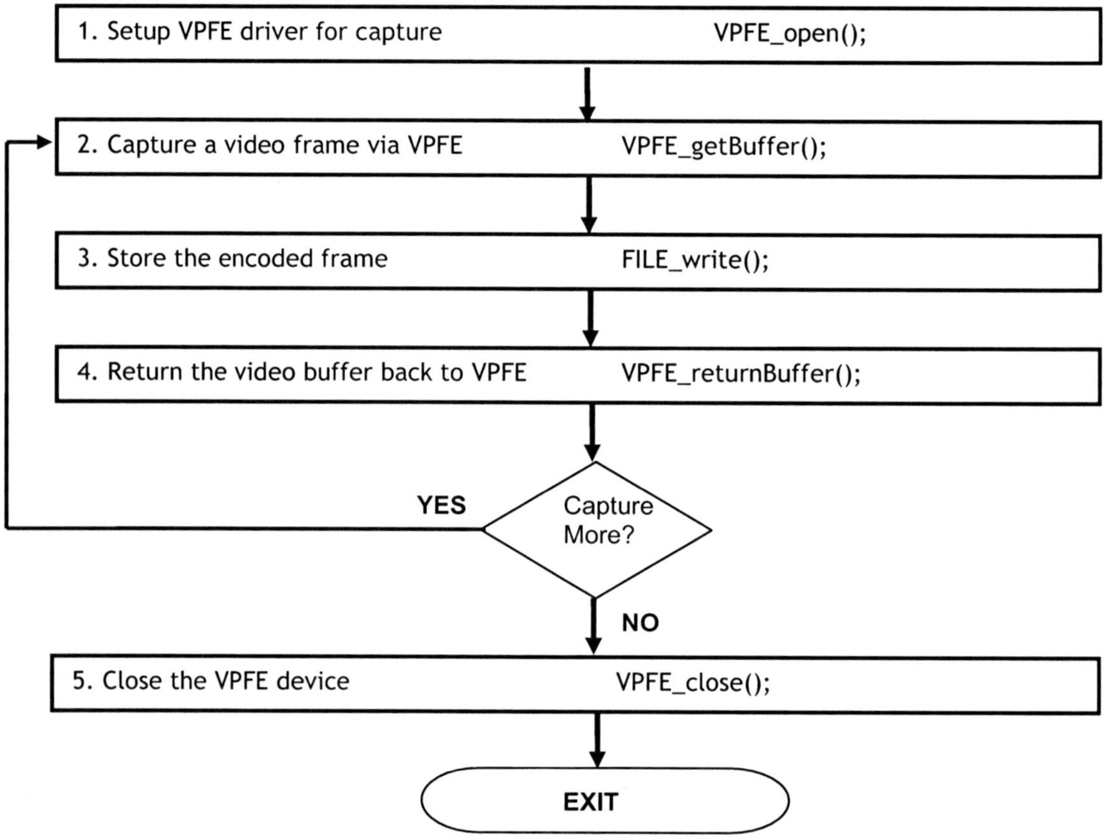

Figure 8.3: Video Capture Pseudo Code.

The actual code for the simple video capture application is shown in the following section. For the sake of readability, initializations and error checking are left out from the code. The application configures the video driver for capturing a 720x480 video frame, captures video on a frame by frame basis and writes the video frame into a file using the file handle provided.

Video Capture Code

```
#define FRAME_SIZE  (720*480*2)

FILE_Handle fileHdl;

VPFE_Params vpfeParams;

void

APP_videoCapture(int numFramesToCapture)

{

  int nframes=0;

  char *frame=NULL;

  /* Initialize VPFE driver */

  vpfeHdl = VPFE_open(&vpfeParams);

  while (nframes++ < numFramesToCapture)

  {

    VPFE_getBuffer(vpfeHdl, frame);

    FILE_write(fileHdl, frame, FRAME_SIZE);

    VPFE_returnBuffer(vpfeHdl, frame);

  }

  VPFE_close(vpfeHdl);

}
```

CHAPTER 9

Sample Application Using EPSI and xDM

This chapter was written by Anand Balagopalakrishnan, Texas Intruments India Pvt. Ltd.

9.1 OVERVIEW

This chapter shows how to write a video encode application using the VISA APIs provided by Codec Engine and EPSI APIs for VPFE. We call this video encode application as a Controller Application since it controls the behavior of both ARM and DSP subsystems using VISA and EPSI APIs. This application builds on the sample application in the earlier chapter titled *Sample Application Using EPSI.*

This chapter also shows how to leverage the video encode application to measure the encoder performance, calculate the Codec Engine latency. It also shows the changes required for running multiple codec instances.

9.2 CONTROLLER APPLICATION DEVELOPMENT

Controller application is a master thread that runs on the ARM/Linux. This application is also referred to as the Digital Video Test Bench (DVTB). More information is located on the web at the Texas Intruments DaVinci & OMAP Developer Wiki (`www.ti.com/davinciwiki_dvtb`). This application controls the flow of data between ATM and DSP, from/to peripherals, between drivers and codecs, etc. In other words, this application acts as a "traffic cop" and directs the flow of data between different components in the system. Different system flows result in different use cases.

An application is built using the services provided by EPSI and VISA APIs. All applications have three stages – creation, execution and deletion.

- ☐ **Creation** – This is the initialization phase. The peripherals and codecs are initialized using the appropriate EPSI and VISA APIs. All components are configured as needed and required resources are allocated.
- ☐ **Execution** – This is the phase where the actual work gets done. Processing of data at each component and chaining of components to ensure a data flow occurs in this stage.
- ☐ **Deletion** – This is the cleanup phase. Once the processing is completed, the allocated resources are freed up. All components configured in the creation phase go through a teardown process.

The flow of a typical DaVinci application is given in Figure 9.1.

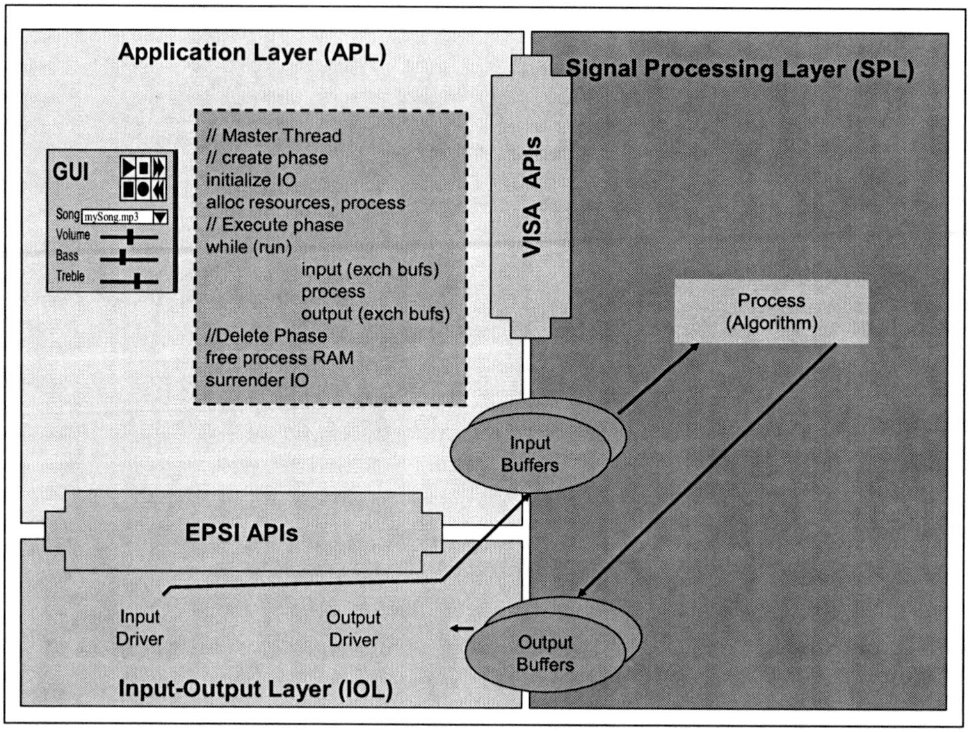

Figure 9.1: Controller Application.

9.3 VIDEO ENCODE APPLICATION

Let us take an example of a video encoder application. This application needs to capture video frames from the Video Processing Front End (VPFE), pass the frames to a video encoder as input and store the encoded frames to a file. Depending on the use case, the application might transmit the encoded frames over the network. For this example, we will consider that the application stores the encoded frames to a file on the hard disk.

This real world example follows the general skeleton of a typical controller application as seen in Section 3.1. Using the VISA and EPSI APIs, we can write this application in just a few lines of code. Figure 9.2 gives the control flow for this application. Figure 9.3 gives the pseudo code for this application.

The following section is the actual C code for this application. For the sake of readability, error checking and parameter initializations have been left out from this code.

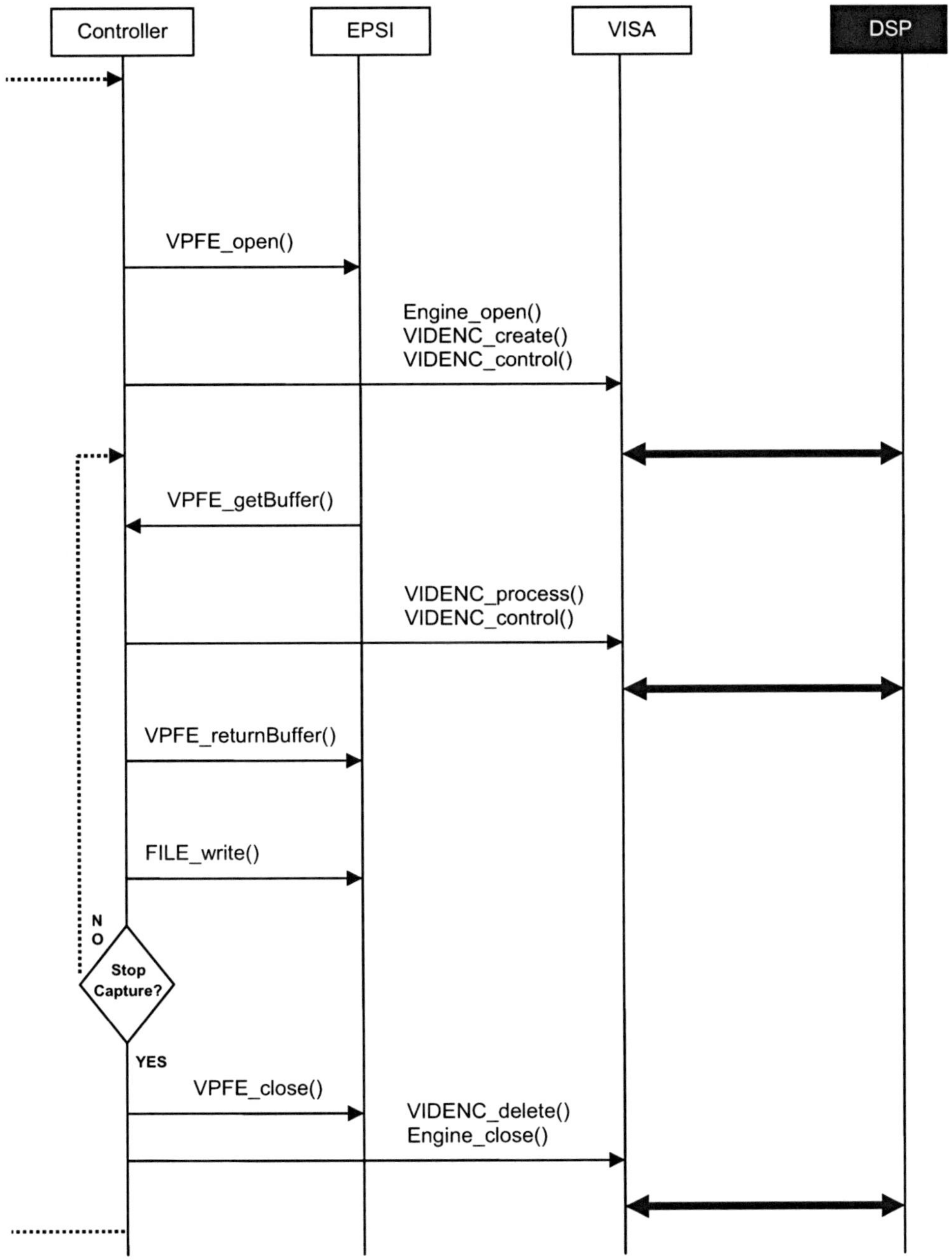

Figure 9.2: Video Encode Control Flow.

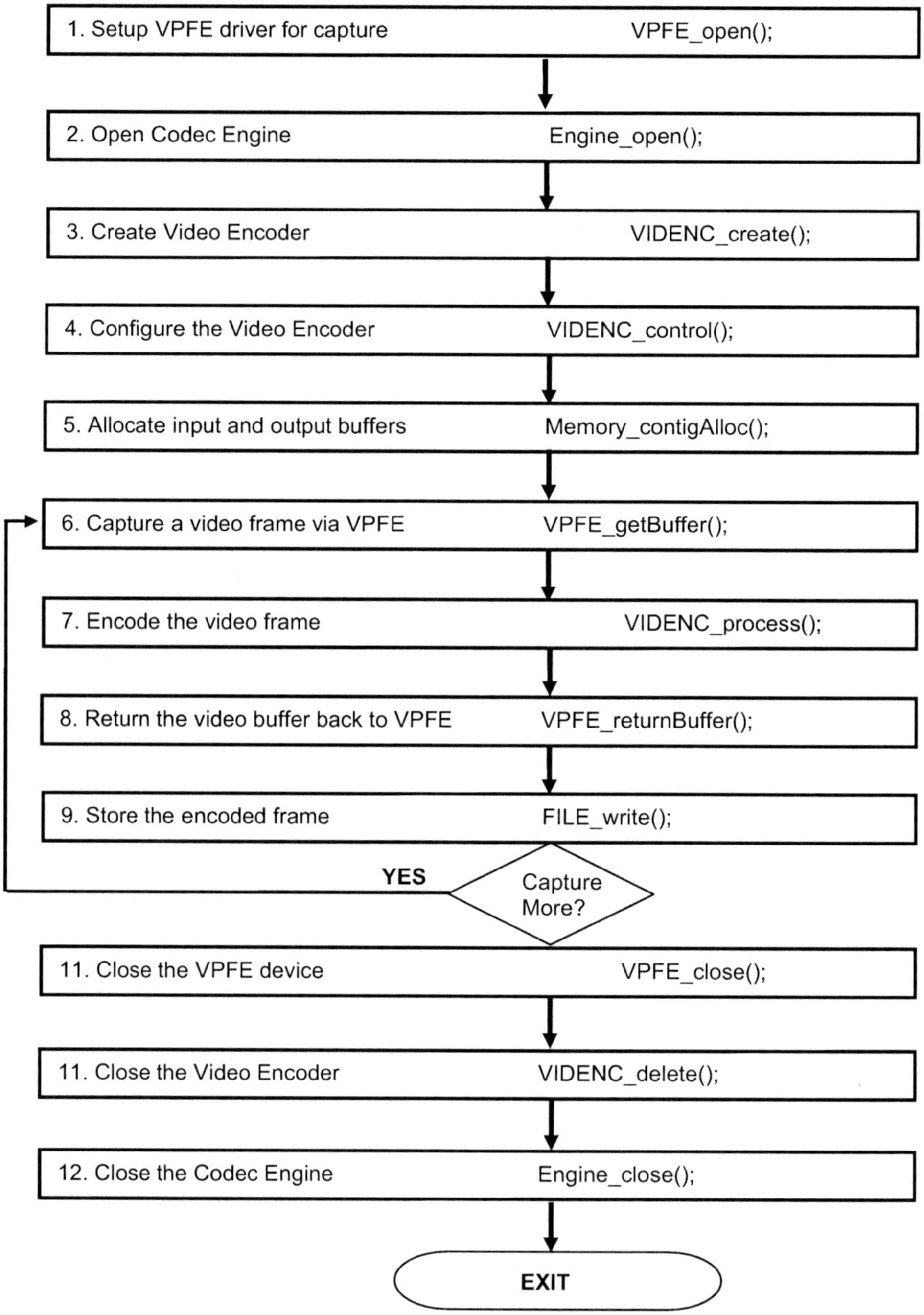

Figure 9.3: Video Encode Pseudo Code.

Video Encode Code

```
void

APP_videoEncode(int numFramesToCapture)

{

  int nframes=0;

  char *frame=NULL;

  /******************************

  * Creation Phase

  *******************************/

 /* Initialize VPFE driver */

  vpfeHdl = VPFE_open(&vpfeParams);

  /* Initialize Codec Engine */

  engineHdl = Engine_open(''encode'', NULL, NULL);

  /* Initialize Video Encoder */

videncHdl = VIDENC_create(engineHdl, ''h264enc'',
                          &videncParams);

  /* Configure Video Encoder */

  VIDENC_control(videncHdl, XDM_SETPARAMS, &videncDynParams,
                 &videncStatus);

 /* Initialize file */
```

```
  fileHdl = FILE_open(``test.enc'', ``w'');

  /******************************

  * Execution Phase

  *******************************/

  while (nframes++ < numFramesToCapture)

  {

    VPFE_getBuffer(vpfeHdl, frame);

    VIDENC_process(videncHdl, &inbufDesc, &outbufDesc,
                           &inArgs, &outArgs);

    VPFE_returnBuffer(vpfeHdl, frame);

    FILE_write(fileHdl, outBufDesc.bufs[0],
               outArgs.bytesGenerated);

  }

  /******************************

  * Deletion Phase

  *******************************/

  VPFE_close(vpfeHdl);

  VIDENC_delete(videncHdl);

  Engine_close(engineHdl);

  FILE_close(fileHdl);

}
```

The application code shown above follows a simple, clean, easy to follow logic. Usage of EPSI and VISA APIs make the application code modular, portable, and readable.

9.4 LEVERAGING THE APPLICATION

This section describes how the video encode application can be leveraged by adding additional features.

9.4.1 PERFORMANCE MEASUREMENTS

The first stage in any software development is to ensure functionality. After verifying functionality, the next step is to determine the system performance and optimize it.

When measuring the system performance, an application developer would want to know the performance of individual components, overheads added by system frameworks and the actual CPU cycles occupied by the application. Out of these, the application developer has a direct control over the application performance. This section describes the tools provided by DaVinci software framework to measure component performance and framework overheads.

The critical performance numbers in the video encode application are the encoder performance itself and the Codec Engine latency. The VISA API `VIDENC_process()` is called on the ARM and the encoding happens on the DSP. Till the DSP completes encoding of the frame, the application blocks on the `VIDENC_process()` function.

The following subsections describe the various methods an application developer can use to get the performance numbers.

9.4.1.1 Codec Engine API

The Codec Engine provides an API `Engine_getCpuLoad()` to get the DSP CPU load. The return value of this function is a number between 0-100 indicating the average DSP CPU load at that point.

The DSP CPU load is the percentage of time that the DSP is doing some useful work. The DSP/BIOS has an idle task that has the lowest priority. When no higher priority tasks are running, the DSP/BIOS schedules the idle task. The Idle Time is the percentage of time when the DSP is idle i.e., when idle task is running. The Idle Time is calculated over a certain period and, hence, the DSP CPU load calculated is a running average of the DSP performance.

```
DSP CPU Load = ( 100 - (Idle Time) )%
```

Figure 9.4 shows how the `Engine_getCpuLoad()` API can be used to determine the DSP performance. This API is called after every `VIDENC_process()` call. The DSP CPU load returned by the API would have factored in the recent codec encode call. If there are no other tasks scheduled on the DSP (e.g., in a single codec instance execution), over a period of time, the DSP CPU load returned by this API will indicate the average DSP cycles consumed by the codec.

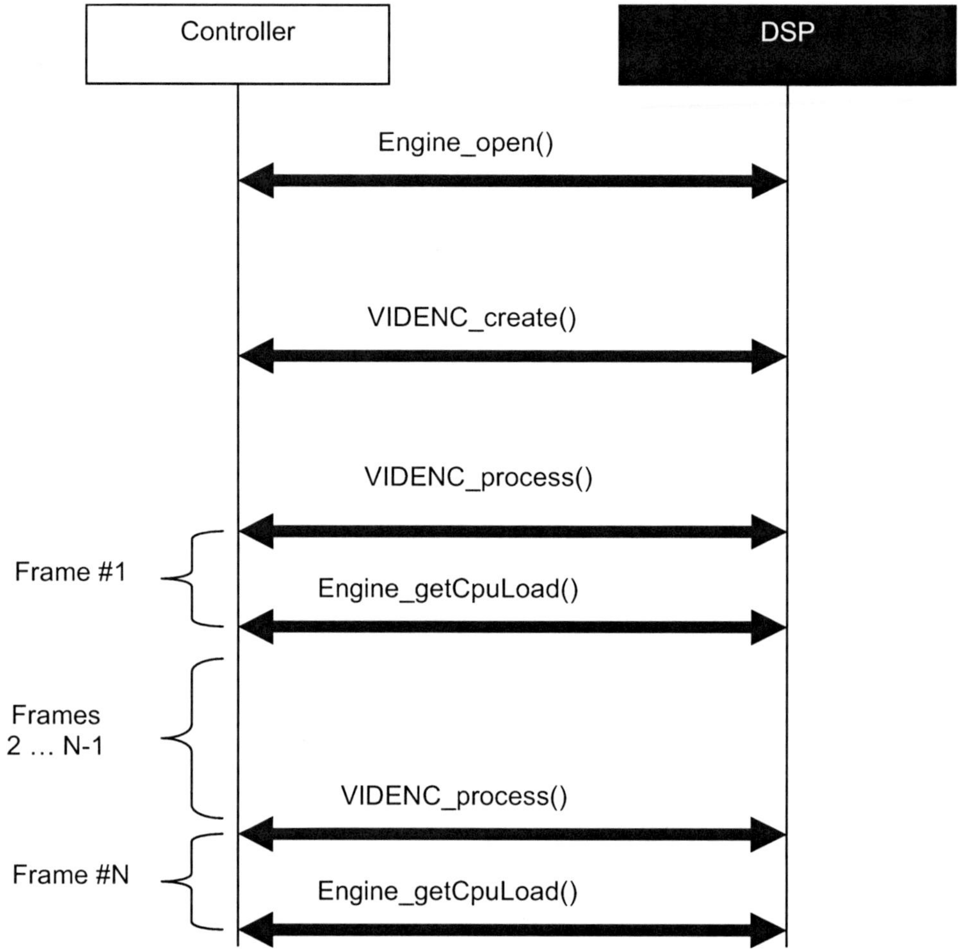

Figure 9.4: Determining DSP performance.

The average DSP CPU load can be calculated as follows:

Let $C(i)$ = DSP CPU load returned by Engine_getCpuLoad for frame ''i''

$$\text{Average DSP CPU Load} = \frac{\sum_{i=1}^{N} C(i)}{N}\%$$

This API lets the application developer determine a quick estimate of the average DSP CPU load in the recent past. It is also useful for monitoring the fluctuations in CPU load.

On the flip side, C(i) is not accurate on a frame by frame basis. For the very first frame, C(1) = 100 since the DSP has been idle till that point. In other words, C(i) takes time to stabilize. Apart from this, the return value is a DSP Engine level CPU load. We cannot isolate the performance from this number. In addition, when two codecs are in operation simultaneously, this method is not reliable.

9.4.1.2 ARM timestamps

This method is the most intuitive to implement in the application. It involves capturing the time-stamps immediately before and after the VISA API `VIDENC_process()` call on the ARM. As seen earlier, the API is a blocking call on ARM and the application unblocks only after the encoding completes on the DSP. The difference between timestamps before and after the `VIDENC_process()` API gives the overall time taken to encode a frame including the actual codec encode cycles, the system framework (CE) latencies, ARM-DSP message passing latencies and any cache maintenance overheads.

Figure 9.5 shows how timestamps can be used to calculate the performance.

After recording the timestamps for each frame, the DSP CPU load is calculated as follows:

```
1. Let's assume that video capture is NTSC. The encode
operation for each frame will occur once in every 33 ms,
i.e., 30 frames are processed in 1 second => each frame
is processed in 33 ms.

2. Let's also assume that the DSP is running at 594 MHz

For frame ''i'':

   B(i) = Time stamp before VIDENC\_process() call

   A(i) = Time stamp after VIDENC\_process() call

   C(i) = A(i) -- B(i) microseconds
```

$$P(i) = \frac{C(i)}{33000} * 594\ \texttt{MHz}$$

$$\texttt{Average Encode Duration} = \frac{\sum_{i=1}^{N} C(i)}{N}\ \texttt{microseconds}$$

$$\texttt{Average DSP MHz} = \frac{\sum_{i=1}^{N} P(i)}{N}\ \texttt{MHz}$$

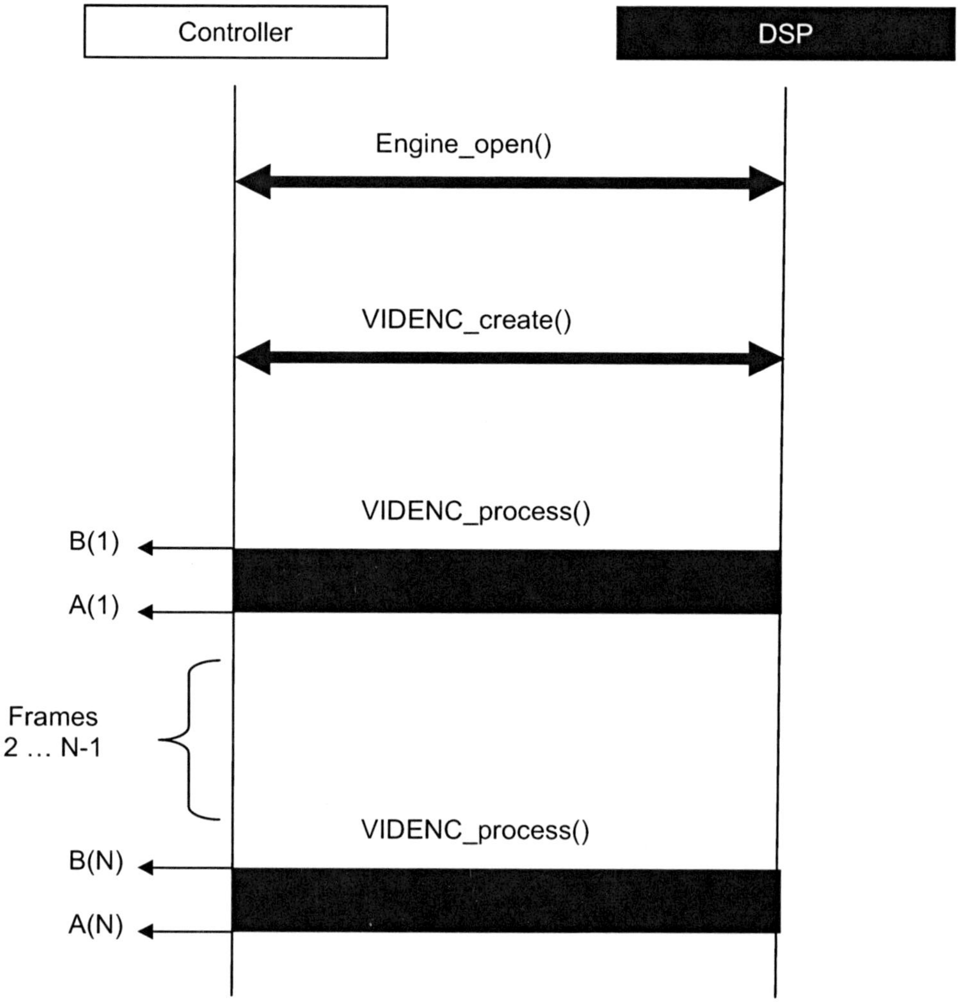

Figure 9.5: ARM Timestamp.

The following code snippet shows how to capture the timestamps on ARM/Linux. The variable "encodeTime" contains the time taken to encode a frame in microseconds. This variable is the same as C(i) above.

Using the formula given above, we can compute the average DSP MHz consumed by the video encode application.

```
#define NUM_MICROSECS_IN_SEC     (1000000)
```

```
typedef struct timeval TimeStamp;

...

  TimeStamp t1, t2;

  int encodeTime;

  /* Get timestamp before and after encode */

   gettimeofday(&t1, 0);

VIDENC_process(videncHdl, &inbufDesc, &outbufDesc,
               &inArgs, &outArgs);

   gettimeofday(&t2, 0);

   /* Calculate the time taken to encode */

   encodeTime = (t2.tv_sec * NUM_MICROSECS_IN_SEC) +
                t2.tv_usec) -
                (t1.tv_sec * NUM_MICROSECS_IN_SEC) + t1.tv_usec)

    ...
```

This method is more granular than using `Engine_getCpuLoad()` described in Section 3.3.1.1. We can get a detailed performance report on a frame-by-frame basis. We can plot and analyze CPU load patterns based on the nature of input content.

On the flip side, there could be an atomicity problem when recording timestamps. For example, if there is a context switch on the ARM side just after the `VIDENC_process()` call but before the timestamp "t2" gets recorded, it will skew the results.

9.4.1.3 Codec Engine Traces

This method uses the trace support provided by Codec Engine to capture the performance data. Refer to the chapter on Codec Engine on the various trace options supported and enabling the traces.

The CE tracing can also be turned on at the time of executing the ARM application by setting the CE trace level from command line. The typical trace level used to capture the performance numbers is trace level 2.

```
root# CE_DEBUG=2 ./controller > trace.txt
```

After traces are turned on, the CE prints detailed Codec Engine traces with timestamps on the console. The trace outputs on the console are captured into a log file. A post processor can be executed on the log file to filter the required CE traces for determining performance.

Figure 9.6 shows a section of sample CE trace when a frame of video is getting encoded.

Figure 9.6 **highlights** in blue the trace lines relevant to performance calculations. The output contains traces from both ARM and DSP.

The ARM – author: see p9-13 - Missing Text?

The different performance numbers can now be derived as follows:

The output contains traces from both ARM and DSP. Each trace line is preceded by the corresponding timestamp. ARM timestamps are recorded in microseconds (us). DSP timestamps are recorded in form of DSP ticks (tk).

The DSP ticks are directly obtained from the TSC registers and right shifted by 8. If <T> is the ticks value printed in the trace line and DSP is running at 594 MHz:
DSP cycles consumed = <T> * 256
Duration in microsecs = (<T> * 256) / 594 us

The important trace lines are marked by a letter in Figure 9.6.
A = Start of CE processing on ARM
B = Start of CE processing on DSP
C = Start of input buffer cache invalidation on DSP
D = End of input buffer cache invalidation on DSP
E = End of algorithm activation on DSP
F = End of codec processing on DSP
G = Start of output buffer cache writeback-invalidate on DSP
H = End of output buffer cache writeback-invalidate on DSP
I = End of CE processing on DSP
J = End of CE processing on ARM

By calculating the difference between appropriate timestamps, different performance numbers can be derived as follows:
J - A = Total time taken to encode a frame from ARM
I - B = Total time taken to encode a frame on DSP
F - E = Actual encoder processing time
D - C = Time taken for input buffer cache invalidation
H - G = Time taken for output buffer cache writeback-invalidate

```
. . .
  @2,386,158us: [+0 T:0x4118eb60] ti.sdo.ce.video.VIDENC - VIDENC_process> Enter
A (handle=0x91af0, inBufs=0x80b2c, outBufs=0x80bfc, inArgs=0x80a0c, outArgs=0x80a18)
  @2,386,331us: [+5 T:0x4118eb60] CV - VISA_allocMsg> Allocating message for
  messageId=0x00020fa6
  @2,386,498us: [+0 T:0x4118eb60] CV - VISA_call(visa=0x91af0, msg=0x4199e880):
  messageId=0x00020fa6, command=0x0
B [DSP] @4,054,353tk: [+5 T:0x8c4cefcc] CN - NODE> 0x8fa9a3e8(h264enc#0)
  call(algHandle=0x8fa9a4a8, msg=0x8fe04880); messageId=0x00020fa6
C [DSP] @4,054,463tk: [+0 T:0x8c4cefcc] OM - Memory_cacheInv> Enter(addr=0x8a02f000,
  sizeInBytes=921600)
  [DSP] @4,055,361tk: [+0 T:0x8c4cefcc] OM - Memory_cacheInv> return
  [DSP] @4,055,413tk: [+0 T:0x8c4cefcc] OM - Memory_cacheInv> Enter(addr=0x89ca8000,
  sizeInBytes=460800)
  [DSP] @4,055,897tk: [+0 T:0x8c4cefcc] OM - Memory_cacheInv> return
  [DSP] @4,055,949tk: [+0 T:0x8c4cefcc] OM - Memory_cacheInv> Enter(addr=0x89e46000,
  sizeInBytes=921600)
D [DSP] @4,056,859tk: [+0 T:0x8c4cefcc] OM - Memory_cacheInv> return
  [DSP] @4,056,912tk: [+0 T:0x8c4cefcc] ti.sdo.ce.video.VIDENC - VIDENC_process> Enter
  (handle=0x8fa9a4a8, inBufs=0x8c4d20e4, outBufs=0x8c4d21b4, inArgs=0x8fe04a04,
  outArgs=0x8fe04a10)
  [DSP] @4,057,036tk: [+5 T:0x8c4cefcc] CV - VISA_enter(visa=0x8fa9a4a8): algHandle =
  0x8fa9a4d0
  [DSP] @4,057,101tk: [+0 T:0x8c4cefcc] ti.sdo.ce.alg.Algorithm - Algorithm_activate>
  Enter(handle=0x8fa9a4d0)
  [DSP] @4,057,173tk: [+0 T:0x8c4cefcc] ti.sdo.fc.dskt2 - _DSKT2_activateAlg> Enter
  (scratchId=0, alg=0x8ba045e8)
  [DSP] @4,057,249tk: [+2 T:0x8c4cefcc] ti.sdo.fc.dskt2 - _DSKT2_activateAlg> Last
  active algorithm 0x8ba045e8, current algorithm to be activated 0x8ba045e8
  [DSP] @4,057,341tk: [+2 T:0x8c4cefcc] ti.sdo.fc.dskt2 - _DSKT2_activateAlg>
  Activation of algorithm 0x8ba045e8 not required, already active
  [DSP] @4,057,422tk: [+0 T:0x8c4cefcc] ti.sdo.fc.dskt2 - _DSKT2_activateAlg> Exit
E [DSP] @4,057,628tk: [+0 T:0x8c4cefcc] ti.sdo.ce.alg.Algorithm - Algorithm_activate>
  return
F [DSP] @4,080,334tk: [+5 T:0x8c4cefcc] CV - VISA_exit(visa=0x8fa9a4a8): algHandle =
  0x8fa9a4d0
  [DSP] @4,080,433tk: [+0 T:0x8c4cefcc] ti.sdo.ce.alg.Algorithm -
  Algorithm_deactivate> Enter(handle=0x8fa9a4d0)
  [DSP] @4,080,661tk: [+0 T:0x8c4cefcc] ti.sdo.fc.dskt2 -  _DSKT2_deactivateAlg> Enter
  (scratchId=0, algHandle=0x8ba045e8)
  [DSP] @4,080,736tk: [+2 T:0x8c4cefcc] ti.sdo.fc.dskt2 - _DSKT2_deactivateAlg> Lazy
  deactivate of algorithm 0x8ba045e8
  [DSP] @4,080,811tk: [+0 T:0x8c4cefcc] ti.sdo.fc.dskt2 - _DSKT2_deactivateAlg> Exit
  [DSP] @4,080,863tk: [+0 T:0x8c4cefcc] ti.sdo.ce.alg.Algorithm -
  Algorithm_deactivate> return
  [DSP] @4,080,923tk: [+0 T:0x8c4cefcc] ti.sdo.ce.video.VIDENC - VIDENC_process> Exit
  (handle=0x8fa9a4a8, retVal=0x0)
G [DSP] @4,081,005tk: [+0 T:0x8c4cefcc] OM - Memory_cacheWb> Enter(addr=0x89e46000,
  sizeInBytes=921600)
  [DSP] @4,081,905tk: [+0 T:0x8c4cefcc] OM - Memory_cacheWb> return
  [DSP] @4,081,957tk: [+0 T:0x8c4cefcc] OM - Memory_cacheWb> Enter(addr=0x8bc183ca,
  sizeInBytes=1207808)
  [DSP] @4,083,134tk: [+0 T:0x8c4cefcc] OM - Memory_cacheWb> return
  [DSP] @4,083,187tk: [+0 T:0x8c4cefcc] OM - Memory_cacheWb> Enter(addr=0x8bd3c680,
  sizeInBytes=603904)
H [DSP] @4,083,796tk: [+0 T:0x8c4cefcc] OM - Memory_cacheWb> return
I [DSP] @4,083,848tk: [+5 T:0x8c4cefcc] CN - NODE> returned from
  call(algHandle=0x8fa9a4a8, msg=0x8fe04880); messageId=0x00020fa6
  @2,404,415us: [+0 T:0x4118eb60] CV - VISA_call Completed: messageId=0x00020fa6,
  command=0x0, return(status=0)
  @2,404,562us: [+5 T:0x4118eb60] CV - VISA_freeMsg(0x91af0, 0x4199e880): Freeing
  message with messageId=0x00020fa6
  @2,404,681us: [+0 T:0x4118eb60] ti.sdo.ce.video.VIDENC - VIDENC_process> Exit
J (handle=0x91af0, retVal=0x0)
. . .
```

Figure 9.6: Sample CE Trace.

This method is the most comprehensive and accurate method to measure the performance of system developed using DaVinci software framework. Using this method, we can calculate the performance of pure codec encode, cache maintenance and overall performance as seen at system level.

On the flip side, this method produces quite a bit of trace messages per frame of encode. Hence, it may not be suitable for real time systems. It will be most useful when the application developer finds a bottle neck in the system and wants to fine tune the different parts of system to optimize performance.

9.4.2 MEASURING THE CODEC ENGINE LATENCY

When the VISA API is called on the ARM, the Codec Engine on them ARM marshals the parameters into a message and sends the message to DSP using DSP Link. The Codec Engine component on DSP receives the message, un-marshals the parameters, activates the appropriate algorithm instance and calls the corresponding codec `_process()` function. Once the `_process()` function is completed in the codec, the CE performs the reverse process now. The results are marshaled into a message and sent over the DSP Link. The CE on ARM receives the message and passes it back to the application. Only now, the controller application on the ARM unblocks from the `VIDENC_process()` call.

In the above sequence, the CE performs a lot of operations in the background – message passing, algorithm activation, cache maintenance. These operations are necessary but introduce latency. This latency will vary depending on the codec and input parameters. For example, if an input of D1 size is passed to the encoder, cache maintenance will take a longer time than if an input of CIF size is passed. Additionally, different codecs have different buffer requirements.

Once we have the performance numbers as shown in Section 3.3.1.3, further performance metrics can be derived as follows:

```
Overheads on DSP = (I-B) - (F-E)
ARM <--> DSP Buffer passing latency = (J-A) - (I-B)
```

Some of the overhead like cache maintenance and algorithm activation are necessary. However, the knowledge of these overheads will enable the application developer to determine the headroom available on DSP. In addition, the application developer can also fine tune the codec configurations depending on how the overheads get affected due to configurations.

The performance report obtained through the CE trace is given in file below. The performance numbers were captured on a per-frame basis. From this data, the codec performance on DSP, overhead on DSP, overall performance on DSP, overall performance as seen from ARM were derived.

The performance report in Figure 9.7 captures data for 1800 frames. Figure 9.8 shows the line graph for the same report for the first 250 frames. This graph is useful for the ready interpretations it provides:

Note: This is an url to an Excel sheet.

http://www.morganclaypool.com/page/pawate

Figure 9.7: Encoder Performance Report.

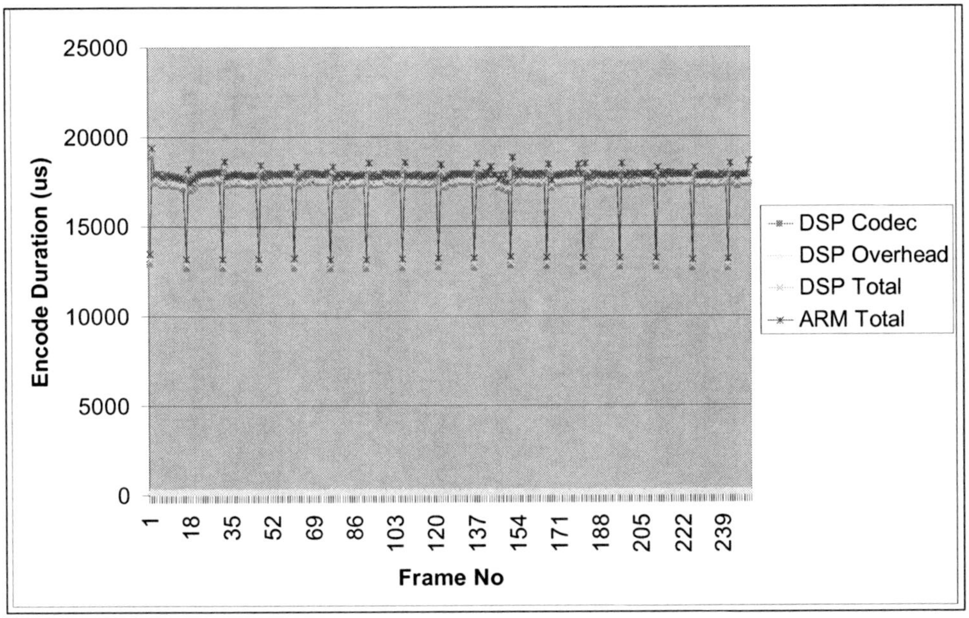

Figure 9.8: Figure Encoder Performance Graph.

- ☐ DSP overhead is constant across frames
- ☐ There are no spikes in the performance graph except at the beginning. This implies that the peak performance will not deviate much from the average performance
- ☐ Total DSP encode time and Total ARM encode time follow each other. This implies that CE latency remains consistent
- ☐ There is a regular trough for every 15 frames. This is understandable as the I-frame interval configured for this test is 15. As the cycles consumed for an I-frame is less than a P-frame, the troughs are seen at regular intervals.

9.4.3 MULTI-CHANNEL APPLICATION

Developing a multi-channel application is the same as writing a single channel application. The only restriction from the Codec Engine is that the Engine handles need to be serialized. This will not be a problem if all the codec instances access the engine handle from the same thread. If the codec instances are running in different instances, then each instance needs to have a separate Engine handle created using the `Engine_open()` API.

We saw the single channel application in Section 3.2. The multi-channel application is provided below. This application opens two input files – test1.yuv and test2.yuv, creates two video encoder instances, and configures both the instances. After the creation phase, this application reads the first input file test1.yuv for the video frame to be encoded, and it passes this frame to the first encoder instance. It reads the second input file test2.yuv, and then it passes the video frame to the second instance of encoder. The encoded frames from instances 1 and 2 are stored into two encoded files test1.enc and test2.enc.

Multi Channel Application Code

```
void

APP_videoEncode(int numFramesToCapture)

{
. . .

/* Initialize Video Encoder instance #1 */

videncHdl1 = VIDENC_create(engineHdl, ''h264enc'',
                                  &videncParams1);

/* Initialize Video Encoder instance #1 */

videncHdl2 = VIDENC_create(engineHdl, ''h264enc'', &videncParams2);

/* Configure Video Encoders */

VIDENC_control(videncHdl1, XDM_SETPARAMS,
               &videncDynParams1, &videncStatus1);

VIDENC_control(videncHdl2, XDM_SETPARAMS,
               &videncDynParams2, &videncStatus2);

/* Initialize file */
```

```
fileIn1 = FILE_open(''test1.yuv'', ''r'');

fileIn2 = FILE_open(''test2.yuv'', ''r'');

fileOut1 = FILE_open(''test1.enc'', ''w'');

fileOut2 = FILE_open(''test2.enc'', ''w'');

while (nframes++ < numFramesToCapture)

{

   FILE_read(fileIn1, inbuf1, FRAME_SIZE);

   FILE_read(fileIn2, inbuf2, FRAME_SIZE);

   VIDENC_process(videncHdl1, &inbufDesc1,
                  &outbufDesc1, &inArgs1, &outArgs1);

   VIDENC_process(videncHdl2, &inbufDesc2, &outbufDesc2,
                  &inArgs2, &outArgs2);

   FILE_write(fileOut1, outbuf1, outArgs1.bytesGenerated);

   FILE_write(fileOut2, outbuf2, outArgs2.bytesGenerated);

}

VIDENC_delete(videncHdl1);

VIDENC_delete(videncHdl2);

Engine_close(engineHdl);

. . .

}
```

CHAPTER 10

IP Network Camera on DM355 Using TI Software Platform

10.1 INTRODUCTION

This document provides detailed information on the source code organization, execution and suggestions to modify ARM and iMX programs on a DM355 IPNetCam Reference Design. DM355 is a multimedia processor from Texas Instruments (TI) with ARM and hardware video accelerator for MPEG4 and JPEG and a set of peripherals for multimedia products. DM355 can support a range of resolutions from SIF to 720p. It can support single as well as multiple channels of MPEG4. The IPNetCam takes input from CMOS Sensors, processes/compresses the video and streams the processed/compressed video over Ethernet. Its web based management console facilitates users for various settings and streaming of video and audio data. Recently, the next generation version of this reference design, based on the DM365 is now available on the TI web at `www.ti.com/ipcamera`

10.2 SYSTEM OVERVIEW

The figure below shows the top-level software architecture of IPNetCam. The IPNetCam software is built on top of the TI DVSDK. The identified partner will work mostly work on the Application Layer (APL) and the Input Output Layer (IOL). The IPNetCam software will use the existing DM365 Codec Engine and provide necessary codec combinations to the users.

This comprises of MontaVista Linux Pro, which is designed for IPNC board using the standard DVSDK Linux kernel. It has various device drivers to support the various interfaces. The application uses this layer with EPSI (Embedded Peripheral Software Interface).

10.3 OPERATING SYSTEM

We will use the MontaVista Linux Pro kernel, which is shipped along with DM365 EVM. This innovative Embedded Linux solution features dynamic power management, rapid kernel boot time, enhanced file systems, new development tools for system performance tuning, and rich processor and peripheral support.

MontaVista Linux comes with TI's xDM VISA APIs. It will be the most efficient and handy to build this solution.

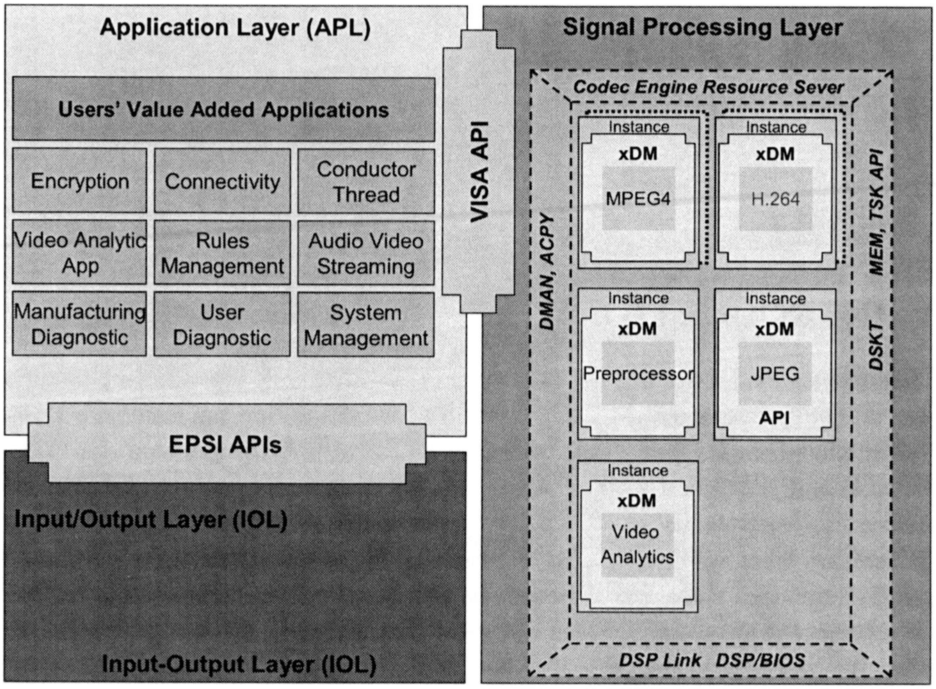

Figure 10.1: IP network camera built on top of DVSDK.

10.4 DEVICE DRIVERS

The IPNetCam has various interfaces, such as Video capture, Audio capture, SD, USB etc. To support all these interfaces, corresponding device drivers for the MontaVista platform needs to be developed/configured. For most of the interfaces, the MontaVista Linux provides the basic driver, which is customized for the actual hardware interface used.

10.5 SUPPORTED SERVICES AND FEATURES

The IPNetCam supports the features that appear in the Tables 10.3, 10.4, and 10.5.

10.6 ACRONYMS

Acronyms are used throughout this document appear in Table 10.6.

10.7 ASSUMPTIONS AND DEPENDENCIES

- This document is based upon the IPNC reference design set with DM355 EVM.

Table 10.3: Application Layer

Connectivity	HTTP Web Server (HTTP) TSL/SSL FTP Server SMTP client NTP client DHCP client UPnP client Network discovery PoE
Audio Video Streaming	Web based video streaming using QuickTime/RealPlayer/VLC to ensure compliance. Additionally, low-latency video player is required on Host PC in order to meet end-to-end latency of 150ms. Audio/video capture date and time are marked up on top of video and inserted in audio/video stream. Audio volume control Play voice alert RTP, RTSP over TCP or UDP
System Management	Multiple user access levels with password protection Firmware Updates on the IPNetCam for further software updates Firmware backup and restore SD and Network Storage settings End-end low latency requirement: 150 ms Analog output for local preview and monitoring the captured (compressed) video/image USB for network detection and configuration
Storage	Local Encrypted Local Storage (MPEG4 SIF stream, JPEG image and alert details can be stored locally.) Network Storage (MPEG4 HD video stream can be stored on a host PC; JPEG and alert details as an email attachment or by FTP protocol.)
Motion detection	Basic motion detection for area of Interest is expected
Manufacturing Diagnostic	A detailed hardware diagnostic software for customers who wants to take this reference design to manufacturing
User Diagnostic	Simple hardware diagnostic software tool to test the basic IO functionality of all peripherals.

Table 10.3: Application Layer (continued)

Image Control	H2A software for auto white balance and auto exposure Zooming of an image using the digital zoom based on ePTZ. This is based on video capture driver; a capture region can be changed frame by frame. By changing the location of capture region, the active scope of the video is Paned/titled/zoomed electrically. IPNC is expected to enable ePTZ at D1 and VGA resolution at 30fps. User-defined video image capture size Switch day/night mode Switch indoor/outdoor mode
Command Control For Admin	Video and Audio channel start & stop ePTZ control (PTZ at a certain step.) Video input setting (720P raw RGB data generating from sensor) Brightness, contrast, saturation, hue, gain setting Setting JPEG parameters (QP) Setting MPEG-4 parameters (CBR/VBR, bitrate, GOP..etc) Setting G.711 parameters Setting Dual Codec Combos (CBR/VBR, bitrate, GOP..etc) Setting area of Interest(ROI) for motion detection Event notification Network Settings User Access control JPEG image storage options Secondary MEPG4 SIF storage options Active Connection List Control the alarm output of the I/O port on the camera Play an audio file(voice alert) Switch day/night mode Switch indoor/outdoor mode Synchronize the date and time of the camera with those of the computer

Table 10.4: Input-Output Layer (IOL)

Video Input	Video Input Driver Video capture directly from CMOS image sensor Video input driver can change capture region and location frame by frame. Enable ePTZ functionality in video driver level (enable ePTZ feature at D1 and VGA resolution at 30fps.) Auto focus, iris, white balancing, dark frame-subtraction, exposure, Lens shading correction using DM355 ISP/VPSS capabilities.
Audio	Mono Input Driver Stereo Output Driver
Storing	SD Memory driver
LAN	EMAC driver POE
GPIO & PWM	GPIO driver
RTC	RTC driver
NAND Flash	NAND Flash driver

Table 10.5: Signal Processing Layer (SPL)

CODEC Combos	MPEG-4 (SP, 720P)+ JPEG Compression + motion detection+G.711 speech codec
Dual Stream CODEC Combo	MPEG4 (SP, 720P) +MPEG4 (SP, SIF) or JPEG (SIF) + motion detection + G.711 speech coding
Triple Stream CODEC Combo	MPEG4 (SP, 720P) +JPEG(VGA)+MPEG4 (SP, SIF) + motion detection + G.711 speech coding

Table 10.6: Acronyms

Acronym	Description
IPNC	Internet Protocol Net Camera
EVM	
CMOS	
2A	Auto White Balance and Auto Exposure

- ☐ The operating system comes from MontaVista Linux version 2.6.10.
- ☐ Texas Instruments Incorporated provides digital Video Software Development Kit (DVSDK).

- ☐ The Code Composer Studio (CCStudio) version CCS 3.3.38.2 or higher is used for flashing to NAND memory.
- ☐ For the single stream and dual stream mode, framerate achieved will be 30 fps.
- ☐ For the triple stream, the MPEG4 (SP, 720P) will be at 30fps, whereas JPEG(VGA) and MPE4 (SP, SIF) will be at 15fps.
- ☐ Motion detection will reduce the frame rate apprx by three fps.
- ☐ For latency under 150 ms, the PC must meet the following requirement.

Hardware

- ☐ Intel(R) Pentium(R) D(DUAL Core) CPU 3.0GHz or equivalent
- ☐ 512 MB system memory or above
- ☐ Sound Card : DirectX 9.0c compatible sound card
- ☐ Video Card : 3D hardware Accelerator Card Required – 100% DirectX 9.0c compatible
- ☐ Ethernet network Port/Card
- ☐ Network Cable
- ☐ 10/100 Ethernet switch/hub

Software

- ☐ VLC media player 0.8.6b or above
- ☐ Windows XP Service Pack 2 or above
- ☐ Resolution of screen setting : 1280x960 or higher for the display of 720P

10.8 SOURCE CODE ORGANIZATION

We now discuss the development tools you need in order to compile the code followed by a brief description of how we have organized the code.

10.8.1 DEVELOPMENT TOOLS ENVIRONMENT(S)

Before starting to build the source code, please ensure that the required software package and building tools is installed correctly. Below is the list for required software:

1. TI DVSDK software package version 1.30.00.40.
2. MontaVista Linux Pro v4.0.1.
3. Root file system for development (optional).

10.8.2 INSTALLATION AND GETTING STARTED

1. Copy <release>/source/ipnc_app_XXXX.tgz into "<installDir>/" directory in your Linux desktop.

2. Uncompress the install file using command below:

```
tar zxvf ipnc_app_XXXX.tgz
```

Then in <installDir>, it will create a directory "ipnc_app/" and a file "Rules.make" and following sub-directories are created. Details of directory structure are as follows:

1. ipnc_app/multimedia/encode_stream/ : Single/Dual Codec/Dual size streaming.
2. ipnc_app/sys_adm/alarm_control/: Demo code for communication with alarm server.
3. ipnc_app/sys_adm/alarm_server/: Alarm server for processing events when event trigger.
4. ipnc_app/sys_adm/file_mng/: Manager for the system parameter.
5. ipnc_app/sys_adm/param_transfer/: Communication interface with web server.
6. ipnc _app/sys_adm/system_control/: Demo code for the communication with system server.
7. ipnc _app/sys_adm/system_control/: Application for processing command from web server.
8. ipnc _app/util/: Common utilities for the process communication.
9. ipnc_app/include/: Common header files.
10. ipnc_app/lib/: Common libraries.
11. ipnc_app/network/boa:
    - ActiveX control of the web server.
    - Display 720P or CIF image on the web browser.
    - Network configuration on the web browser.
12. ipnc_app/network/live: Adding getting CIF image by RTP.
13. ipnc_app/network/msmtp-1.4.13/: Message e-mail sender.
14. ipnc_app/network/quftp-1.0.7/: FTP client for sending jpg image periodically.
15. ipnc _app/network/WebData/: Homepage and some data for web server to use.

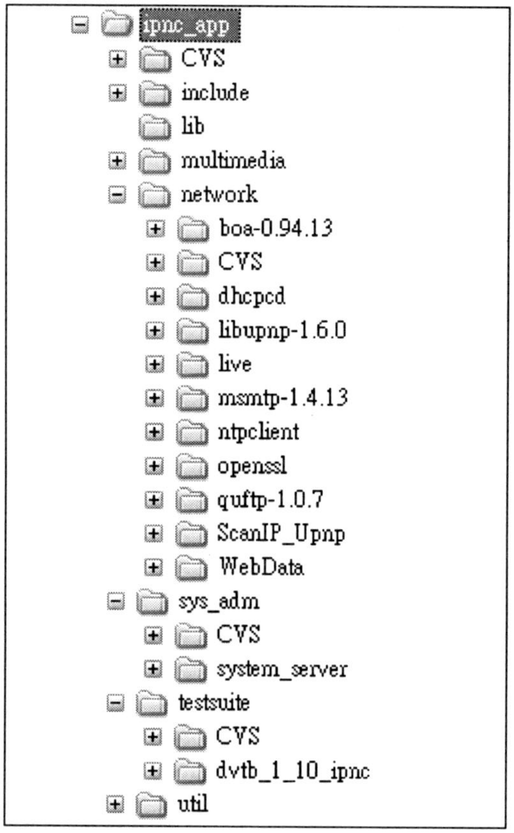

Figure 10.2: Directory structure of IP Netcam software.

3. Once the installation is complete, one needs to modify "Rules.make" based on the system's deployment. Shown below is a simple description about "Rules.make" for reference. Please set the correct environment paths on your system:

   # The installation directory of the DVSDK dvsdk_1_30_00_23.

   DVSDK_INSTALL_DIR=/home/user/workdir/dvsdk_1_30_00_23

   # For backwards compatibility.

   DVEVM_INSTALL_DIR=$(DVSDK_INSTALL_DIR)

   # Where the Codec Engine package is installed.

   CE_INSTALL_DIR=$(DVSDK_INSTALL_DIR)/codec_engine_2_00

   # Where the XDAIS package is installed.

   XDAIS_INSTALL_DIR=$(DVSDK_INSTALL_DIR)/xdais_6_00

```
# Where the DSP Link package is installed.
#LINK_INSTALL_DIR=$(DVSDK_INSTALL_DIR)/NOT_USED
# Where the CMEM (contiguous memory allocator) package is installed.
CMEM_INSTALL_DIR=$(DVSDK_INSTALL_DIR)/cmem_2_00
# Where the codec servers are installed.
CODEC_INSTALL_DIR=$(DVSDK_INSTALL_DIR)/dm355_codecs_1_06_01
# Where the RTSC tools package is installed.
XDC_INSTALL_DIR=$(DVSDK_INSTALL_DIR)/xdc_3_00_02_11
# Where Framework Components product is installed.
FC_INSTALL_DIR=$(DVSDK_INSTALL_DIR)/framework_components_2_00
# Where DSP/BIOS is installed.
BIOS_INSTALL_DIR=$(DVSDK_INSTALL_DIR)/
# The directory that points to your kernel source directory.
LINUXKERNEL_INSTALL_DIR=/home/user/workdir/ti-davinci
# The prefix to be added before the GNU compiler tools (optionally including
# path), i.e., "arm_v5t_le-" or "/opt/bin/arm_v5t_le-".
MVTOOL_DIR=/opt/mv_pro_4.0.1/montavista/pro/devkit/arm/v5t_le
MVTOOL_PREFIX=$(MVTOOL_DIR)/bin/arm_v5t_le-
# Where to copy the resulting executables and data to (when executing 'make
# install') in a proper file structure. This EXEC_DIR should either be visible
# from the target, or one will have to copy this (whole) directory onto
the
# target filesystem.
EXEC_DIR=/home/user/workdir/filesys/opt/net
# The directory that points to the IPNC software package
IPNC_DIR=/home/user/workdir/ipnc_app
# The directory to application include
PUBLIC_INCLUDE_DIR=$(IPNC_DIR)/include
# The directory to application library
LIB_DIR=$(IPNC_DIR)/lib
# The directory to root directory of your root file system
ROOT_FILE_SYS = /home/user/workdir/filesys
```

4. If the login is not as root, one needs to use below commands to prevent error when installation

   ```
   chown -R <useracct> <IPNC_DIR>
   chown -R <useracct> <ROOT_FILE_SYS>
   ```

Substitute your user name for <useracct> and <IPNC_DIR> and <ROOT_FILE_SYS> is the directories you set at "Rules.make" at Step 3.

10.8.3 LIST OF INSTALLABLE COMPONENTS

Note: Any links appearing on this manifest were verified at the time it was created. TI makes no guarantee that they will remain active in the future.

10.8.4 BUILD PROCEDURE

1. Change directory to the <InstallDir>/ipnc_app/ using below command

   ```
   cd <InstallDir>/ipnc_app/
   ```

2. Build the software package using command

   ```
   make clean
   make
   ```

3. Install the application to your root file system

   ```
   make install
   ```

   Note:
   This installation will overwrite files at /etc /var in your root file system.
   Please backup your data first before you start to run.

10.8.5 EXECUTION PROCEDURE

In order to launch Encode Demo, the following command needs to be executed from the target command prompt:

1. # test command for sensor 640x480

   <target prompt># ./encode_ipnc -t 10 -d -r 640x480 -b 200000 -v record_480P.mpeg4

2. # test command for sensor 1280x720

   <target prompt># ./encode_ipnc -t 10 -d -r 1280x720 -b 200000 -v record_720P.mpeg4

After the build is successful, following modules will be generated at the directory $(EXEC_DIR) set at $(installDir)/Rules.make

1. "wis-streamer" "wis-streamer2" "file_mng"

Filename	License						Source	
File or directory name	TI RDSLA Source code distribution not permitted	GPL Source Code Distribution per the GPL		LGPL Source Code Distribution per LGPL		Other Read and follow applicable license terms	Original source obtained from	Original Source modified by TI? (Y/N)
		v2	v3	v2	v3			
Boa Webserve		✓					http://www.boa.org Version: 0.94.13 downloaded: 03 Aug 2007	Y
Dhcpcd		✓					http://www.phystech.com/download/ version: v.1.3.22-pl4 downloaded: 03 Aug 2007	N
ntpclient		✓					http://doolittle.icarus.com/ntpclient/ Version:2007_365 downloaded: 31 Dec 2007	N
libesmtp		✓					http://www.stafford.uklinux.net/libesmtp/download.html Version 1.0.4 Downloaded 1 Mar 2008	N
Esmtp		✓					http://esmtp.sourceforge.net/download.html Version 0.6.0 Downloaded: 1 Mar 2008	N
Quftp		✓					http://sourceforge.net/projects/quftp Version 1.0.7 Downloaded: 31 Dec 2007	N
Libupnp						✓ BSD (Berkeley Standard Distribution) license See LICENSE file or http://pupnp.sourceforge.net/	http://pupnp.sourceforge.net/ Version: 1.6.0 downloaded: 03 Aug 2007	N
FFMpeg		✓					http://ffmpeg.mplayerhq.hu/legal.html Version: SVN-r12347 downloaded: 31 Dec 2007	Y
LIVE555 Streaming Media				✓			http://www.live555.com/liveMedia/public/ Version:2007.08.03 downloaded: 03 Aug 2007	Y

Figure 10.3: Installed Components.

2. "encode_stream"
3. "test.m4e"
4. "boa"
5. "loadmodules_ipnc.sh"

Before you starting the streaming, ensure below files is at the directory $(EXEC_DIR) set at $(installDir)/Rules.make

1. "dm350mmap.ko"
2. "cmemk.ko"
3. "mapdmaq"

Example for execution:

1. VGA Demo:

Start:

$cd $(EXEC_DIR)
$./encode_stream -u 0 -q 50 -d -r 640x480 -b 4000000 –v test.mpeg4 &
$./wis-streamer &

Leave:

$killall -9 wis-streamer
$killall -9 encode_stream

2. 720P Demo:

Start:

$cd $(EXEC_DIR)
$./encode_stream -u 0 -q 50 -d -r 1280x720 -b 4000000 -v test.mpeg4 &
$./wis-streamer &

Leave:

$killall -9 wis-streamer
$killall -9 encode_stream

3. Dual (720P+CIF) Demo:

Start:

$cd $(EXEC_DIR)
$./encode_stream -u 3 -q 50 -d -r 1280x720 -e 352x192 -b 4000000 -v test.mpeg4 &
$./wis-streamer &
$./wis-streamer2 &

Leave:

$killall -9 wis-streamer
$killall -9 wis-streamer2
$killall -9 encode_stream

10.9 ARM9EJ PROGRAMMING

This section explains the top-level process and threads in the system, task partitioning across MJCP and ARM9EJ for codecs, ARM9EJ load, thread/process scheduling, and component addition or deletion.

10.9.1 ARM9EJ TASK PARTITIONING

10.9.1.1 Process/Threads and Scheduling

There are multi processes within IPNC SW to enable various functions including video capture, compression, streaming and configurations. The most important processes are described below.

Encode_stream: Enable video capture, resizing, compression 2A algorithms and motion detection.

Wis-streamer: Take MPEG4 elementary stream from encoder_stream process, pack it into RTP packets and enable streaming over IP.

Webserver: HTTP server based on BOA. ActiveX control is included to display MPEG4 streaming. Network configuration is supported in this webserver process.

A inter-process interface function (getAVdata()) is implemented to ease process-level synchronization and communication. A part of getAvdata is referred below.

```
int GetAVData( unsigned int field, int serial, AV_DATA * ptr )

{

  int ret=RET_SUCCESS;

  if(virptr == NULL)

  return RET_ERROR_OP;

  switch(field) {

  case AV_OP_GET_MJPEG_SERIAL:

    if(serial != -1) {

 ret = RET_INVALID_PRM;

    } else {
```

```
FrameInfo_t curframe = GetCurrentFrame(FMT_MJPEG);

if(curframe.serial_no < 0) {

  ret = RET_NO_VALID_DATA;

} else {

  ptr->serial = curframe.serial_no;

  ptr->size = curframe.size;

  ptr->width = curframe.width;

  ptr->height = curframe.height;

}

   }

 break;
```

Both Wis-streamer and Websercer reuse well-known open resource. Developers should be able to find enough details online. Encode_stream process is the most important process. We are going to discuss it in detail. Every thread within this process will be addressed.

The encode_stream process consists of nine separate POSIX threads (pthreads): the main thread (main.c), which eventually becomes the control thread (ctrl.c), the video thread (video.c), the display thread (display.c), the capture thread (capture.c), the stream writer thread (writer.c), the 2A thread (appro_aew.c), the motion detection thread (motion_detect.c), the audio/video message thread (stream.c), and the speech thread (speech.c). The video, display, capture, writer, 2A, motion, stream interface, and speech threads are spawned from the main thread before the main thread becomes the control thread. All threads except the original main/control thread are configured to be preemptive and priority-based scheduled (SCHED_FIFO). The video and 2A threads share the highest priority, followed by the stream writer, display and capture threads. The speech, and motion thread has lower priority than the writer and capture threads, and the control thread has the lowest priority of all.

The initialization and cleanup of the threads are synchronized using the provided Rendezvous utility module. This module uses POSIX conditions to synchronize thread execution. Each thread

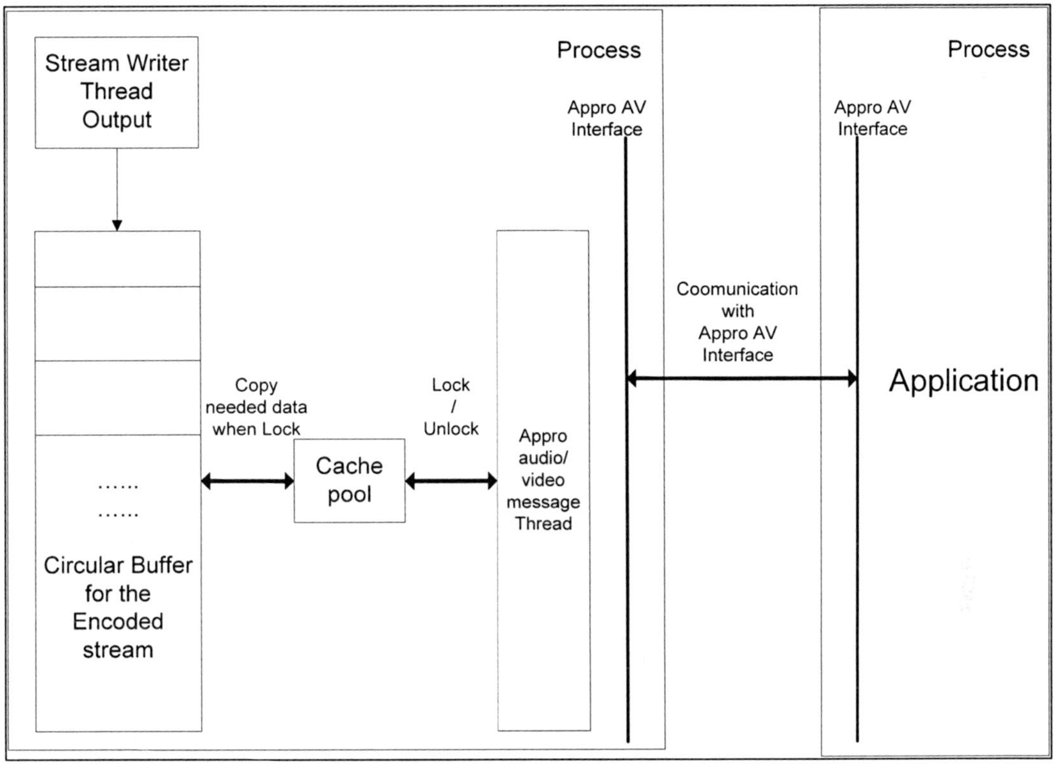

Figure 10.4: Application processes and threads.

performs its initialization and signals the Rendezvous object when completed. When all threads have finished initializing, all threads are unlocked simultaneously and start executing their main loops. The same method is used for thread cleanup. This way buffers that are shared between threads are not freed in one thread while still being used in another.

10.9.1.2 Main Thread

The job of the main thread is to perform necessary initialization tasks, to parse the command-line parameters provided by the user when invoking the application, and to spawn the other threads with parameters depending on the values of the command-line parameters.

10.9.1.3 Display Thread

In order to show a preview of the frames being encoded while they are being encoded, the captured raw frames from the VPSS front end need to be copied to the frame buffer of the VPSS back end. To allow the copying to be performed in parallel with the DSP processing, it is performed by a separate display thread. The thread execution begins by initializing the FBDev display device driver

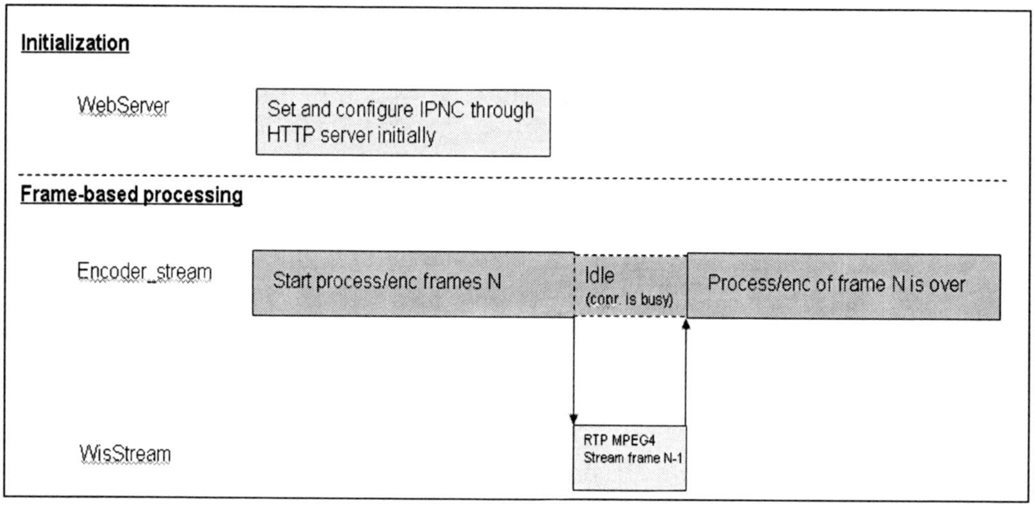

Figure 10.5: Frame based processing of IP Netcam.

in initDisplayDevice(). In this function, the display resolution (D1) and bits per pixel (16) are set using the FBIOPUT_VSCREENINFO ioctl, before the three (triple buffered display) buffers are made available to the user space process from the Linux device driver using the mmap() call. The buffers are initialized to black, since the video resolution might not be full D1 resolution, and the background of a smaller frame should be black. Next, a Rszcopy job is created. The Rszcopy module uses the VPSS resizer module on the DM355 to copy an image from source to destination without consuming CPU cycles. When the display thread has finished initializing, it synchronizes with the other threads using the Rendezvous utility module. Because of this, only after the other threads have finished initializing is the main loop of the display thread executed.

10.9.1.4 Capture Thread

The video capture device is initialized by initCaptureDevice(). The video capture device driver is a Video 4 Linux 2 (v4l2) device driver. In this function, the capabilities of the capture device are verified using the VIDIOC_QUERYCAP ioctl. Next the video standard (NTSC or PAL) is auto-detected from the capture device and verified against the display video standard selected on the Linux kernel command line. Next three video capture buffers are allocated inside the capture device driver using the VIDIOC_REQBUFS ioctl, and these buffers are mapped to the user space application process using mmap(). Finally, the capturing of frames in the capture device driver is started using the VIDIOC_STREAMON ioctl.

10.9.1.5 Stream Writer Thread

To allow the writing of encoded video frames to the circular memory buffer to be done in parallel with the DSP processing, the stream writing is performed by a separate writer thread. First the

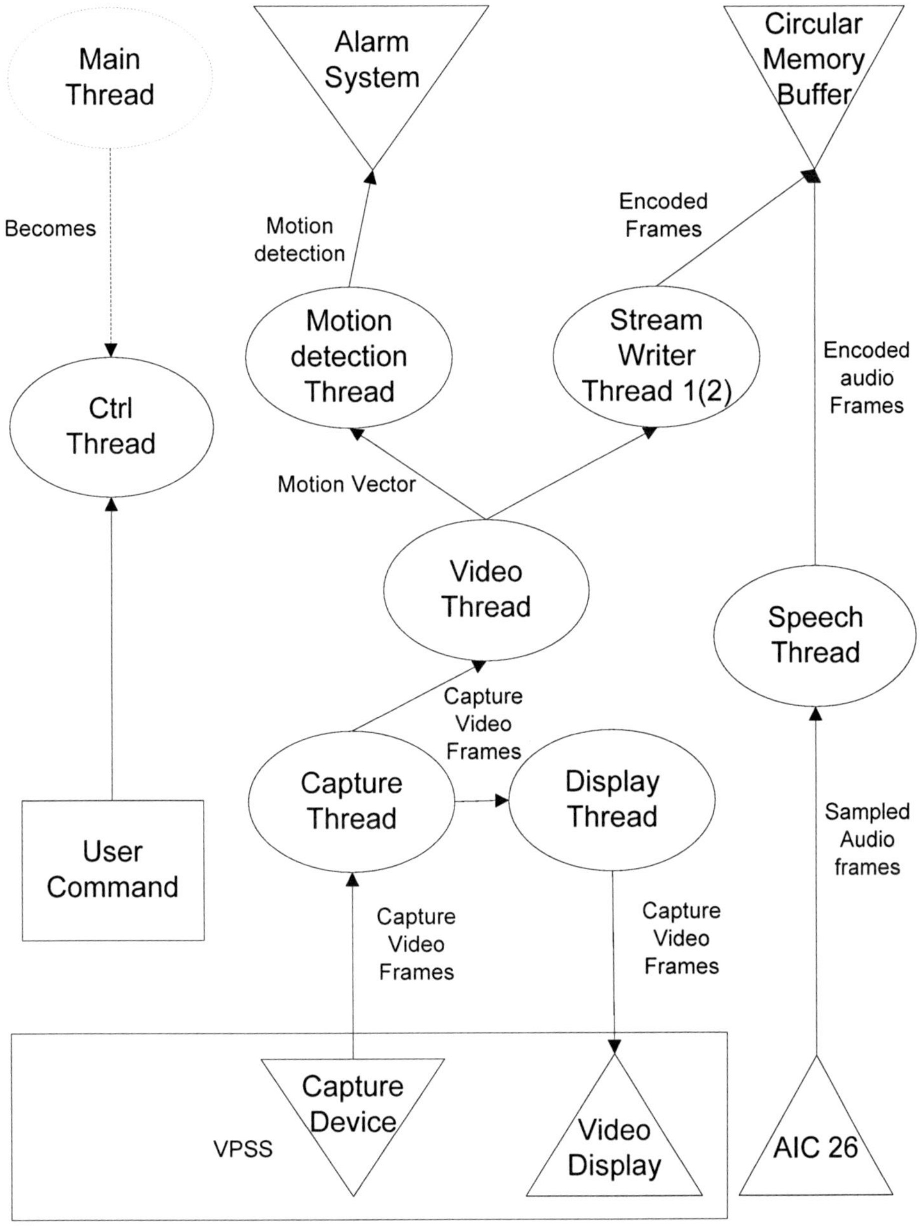

Figure 10.6: Flowchart of thread and user commands.

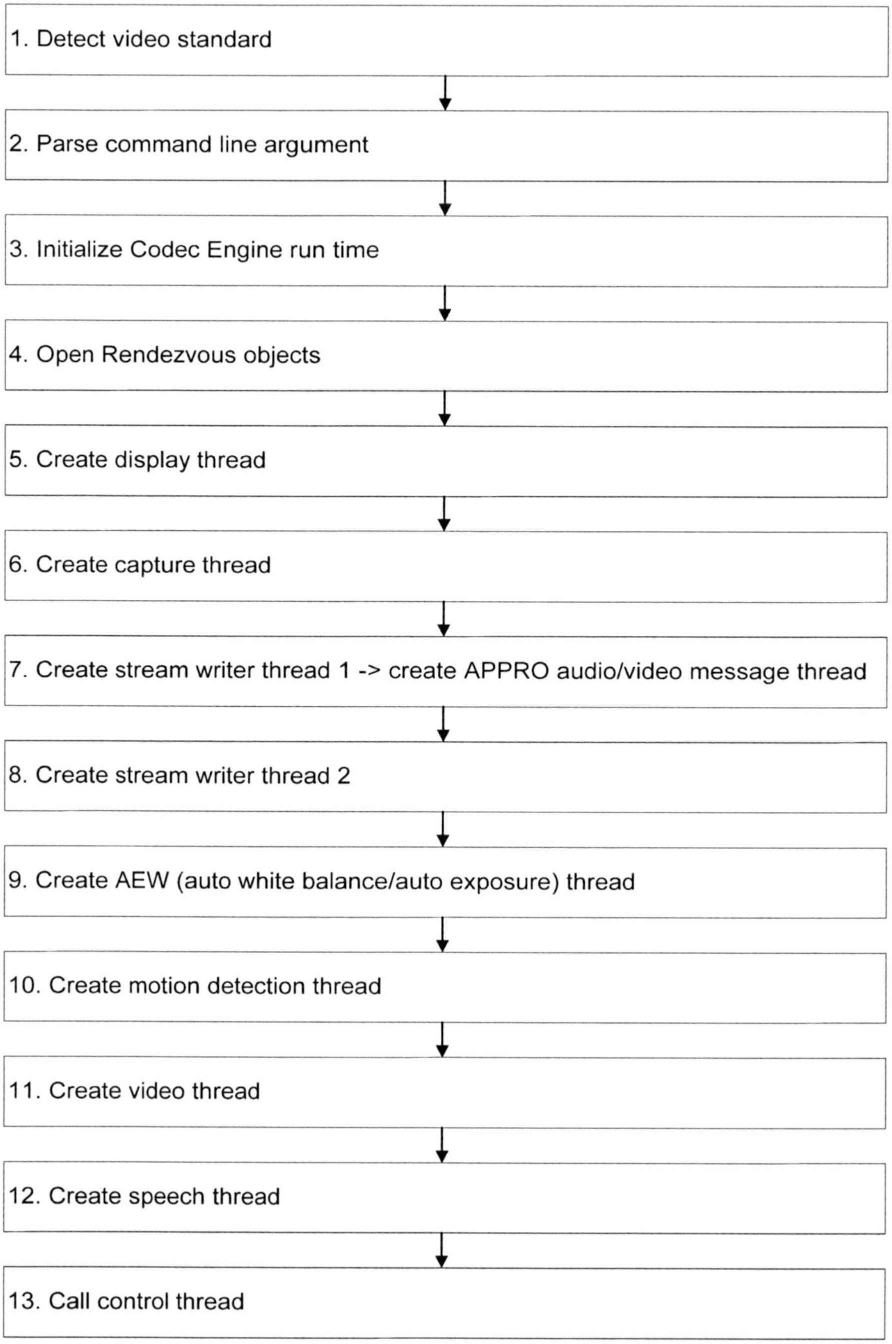

Figure 10.7: Processing sequence.

destination buffer on memory manage is allocated by stream_init (). Then the Rendezvous object is notified that the stream writer thread's initialization is complete. Note that the speech thread, unlike the video thread, does its writing circular buffer in the speech thread itself. This is because speech has lower performance requirements than video.

10.9.1.6 Video Thread Interaction

The figure below shows one iteration of each of the threads involved in processing a video frame once they start executing their main loops, and how these threads interact.

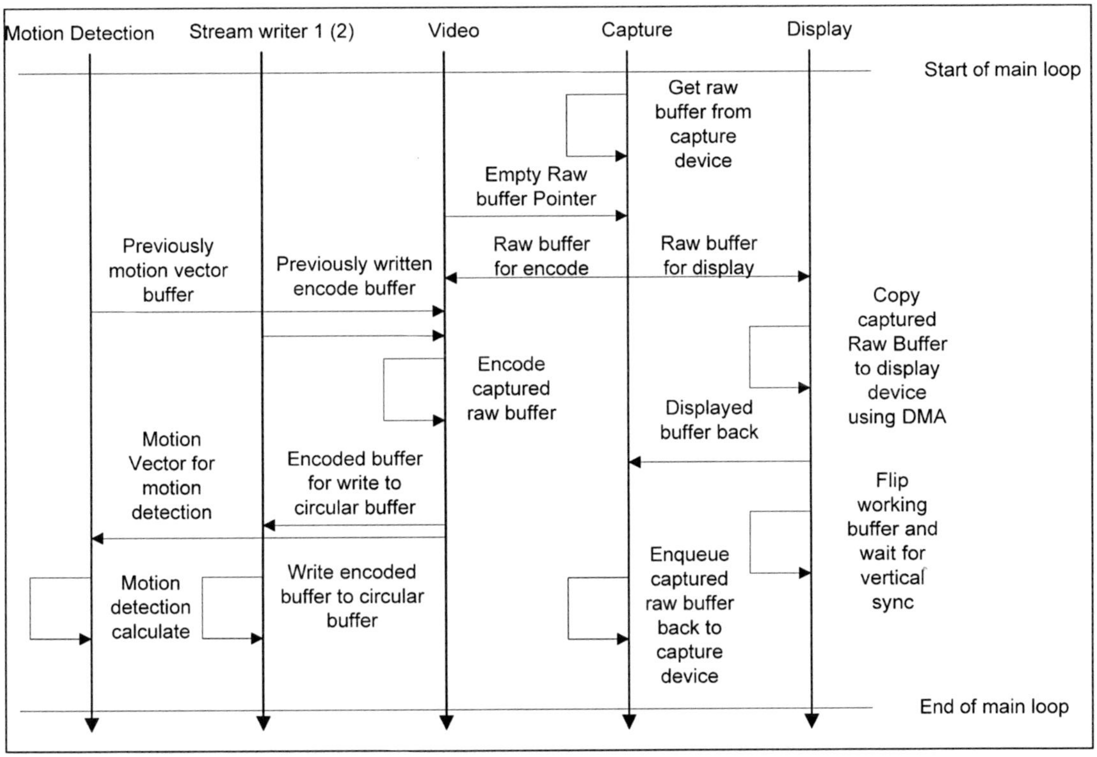

Figure 10.8: More on processing sequence, control, and threads.

First the capture thread dequeues a raw captured buffer from the VPSS front end device driver using the VIDIOC_DQBUF ioctl. To show a preview of the video frame being encoded, a pointer to this captured buffer is sent to the display thread using FifoUtil_put(). The capture thread then fetches an empty raw buffer pointer from the video thread. Then this buffer pointer is then sent to the video thread for encoding.

The video thread receives this captured buffer pointer and then fetches an I/O buffer using FifoUtil_get() from the stream writer thread. The encoded video data will be put in this I/O buffer.

While the display thread copies the captured raw buffer to the FBDev display frame buffer using the Rszcopy_execute() call, the video thread is encoding the same captured buffer into the fetched I/O buffer on the DSP using the VIDENC_process() call. Note that the encoder algorithm on the DSP core and the Rszcopy module might access the captured buffer simultaneously, but only for reading. When the display thread has finished copying the buffer, it makes the new frame buffer to which we just copied our captured frame thenew display buffer on the next vertical sync using the FBIOPAN_DISPLAY ioctl before the thread waits on the next vertical sync using the FBIO_WAITFORVSYNC ioctl. When the video encoder running on the DSP core is finished encoding the captured buffer into the I/O buffer, the I/O buffer is sent to the writer thread using FifoUtil_put(), where it is written to the circular memory buffer using the call stream_write(). The capture raw buffer pointer is sent back to the capture thread to be refilled. The captured buffer pointer is "collected" in the capture thread from the display thread as a "handshake" that indicates that the display copying of this buffer is finished using FifoUtil_get(), before the captured buffer is reenqueued at the VPSS front end device driver using the VIDIOC _QBUF ioctl. The writer thread writes the encoded frame to the circular memory buffer while the capture thread is waiting for the next dequeued buffer from the VPSS front end device driver to be ready. If the writing of the encoded buffer is not complete when the next dequeued buffer is ready and the capture thread is unblocked, there is no wait provided IO_BUFFERS is larger than 1 since another buffer will be available on the FIFO at this time. The encode stream application has IO_BUFFERS set to 2.

10.9.2 ARM CPU UTILIZATION

ARM CPU (running at 216 MHz) utilization is profiled statistically. The CPU loading information is collected 300 times in a period of 5 minutes. The details are listed below.

Note:
ARM is running at 216MHz.

10.10 IMX PROGRAMMING

This section provides ways of offloading computational load to iMX available in DM355, which runs concurrently with ARM9EJ. Special treatment is required as the image and video codecs like JPEG and MPEG4 are tightly coupled to iMX and other coprocessors/accelerators. iMX is free for 70 to 84% of the MPEG4 encoder execution time depending on encoder settings.

10.10.1 IMX PROGRAM EXECUTION

The iMX program runs concurrently with ARM9EJ. Typical iMX program are math intensive requiring MAC operations. iMX in DM355 can perform 4 MACs per cycle. iMX and ARM9 run at same clock (216 MHz on DM355H and 271 MHz on DM355UH).

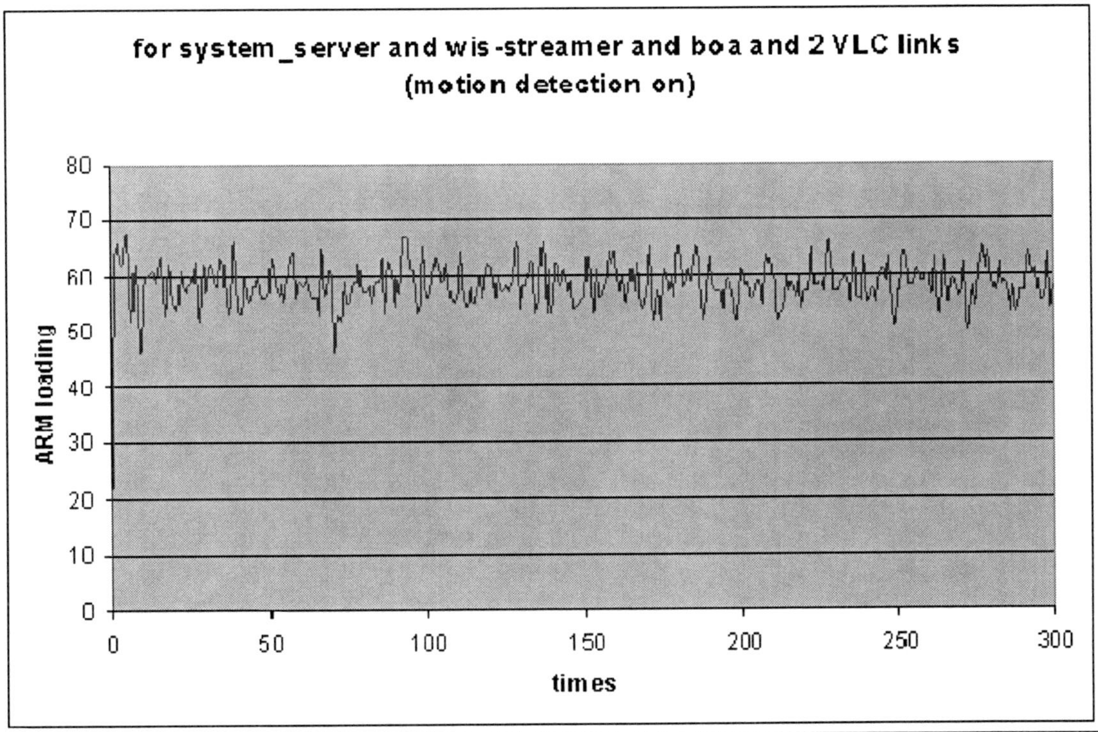

Note: ARM is running at 216MHz

Figure 10.9: Measuring ARM CPU utilization and determining available headroom.

10.10.1.1 iMX Utilization by MPEG4 Encoder

MPEG4 encoder uses iMX for performing color conversion. If 8x8 intra/inter decision is enabled, iMX is used for 8x8 average computation needed for intra/inter decision logic. This is about 400 cycles on iMX.

10.10.1.2 Sequential to MPEG4 Encoder

iMX algorithms can be run sequentially with MPEG4 JPEG Coprocessor (MJCP) as shown in Figure 10.10. In this case, entire SEQ, and iMX program and data memory is available for the algorithm. iMX execution cycles corresponding to unused MJCP cycles is available for algorithms. The feasibility of such scenario depends on MJCP free time. In other words, iMX can run when MJCP is idle. The activate() and deactivate() xDM calls implemented by codecs protect against context switches in iMX and MJCP usage. Similarly, the algorithms will have save iMX context if needed by implementing activate and deactivate calls. Activate is used to restore context and deactivate for saving context.

In addition, the iMX program memory can be extended to 4096 bytes (instead of 1024 byes) by swapping command memory with MJCP (since MJCP is not executing).

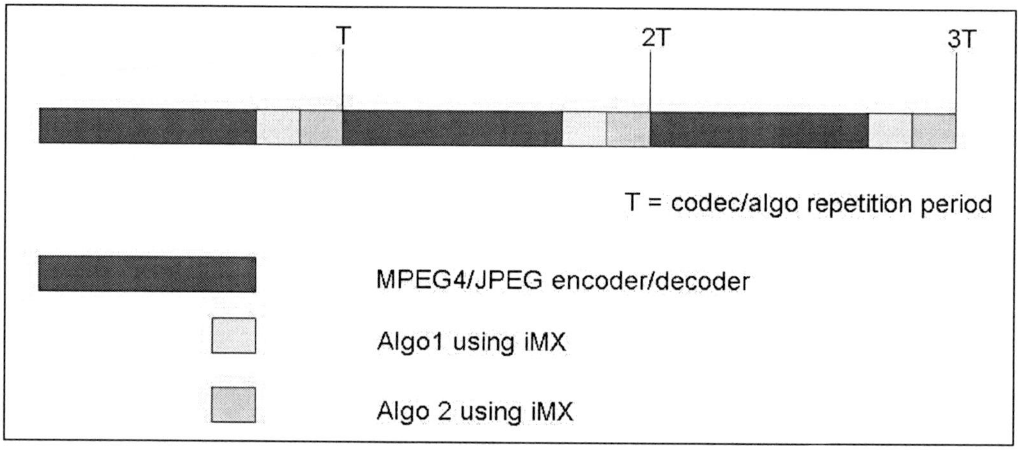

Figure 10.10: Sequential execution of iMX programs with MPEG4/JPEG codecs.

10.10.1.3 Concurrent with MPEG4 Encoder

In case of concurrent execution, iMX programs are executed in parallel with MJCP execution as shown in Figure 10.11.

The availability of iMX for processing other than encode/decode operations depends on:

Availability of IO memory (image buffer and coefficient buffer)

iMX program for algorithms should use the space not used by codecs. 3848 bytes out of 4096 bytes of image buffer is used (248 bytes free in image buffer). 4352 bytes out of 8192 bytes in coefficient buffer is used (3840 bytes free in coefficient buffer).

Availability of program memory (command buffer)

iMX program of algorithm will have to be inserted before or after iMX program for codec. 468 bytes out of 1024 is used (556 bytes free in command memory). iMX program start and end addresses are 0x11F06000 and 0x11F061D4 respectively.

Availability of SEQ memory (program and data)

Sequencer is required for scheduling DMA transfers to fetch data into or out of iMX image and coefficient memory. If the iMX program is included as extension to the codec (either pre- or post-processing operating on same data as the codec) then SEQ_code of codec may not require change to handle extra program in iMX. 2560 bytes out of 4096 bytes in program memory is used (1536 bytes of program memory is free. No data memory). SEQ_program start and end addresses are 0x11F0F000 and 0x11F0FA00 respectively.

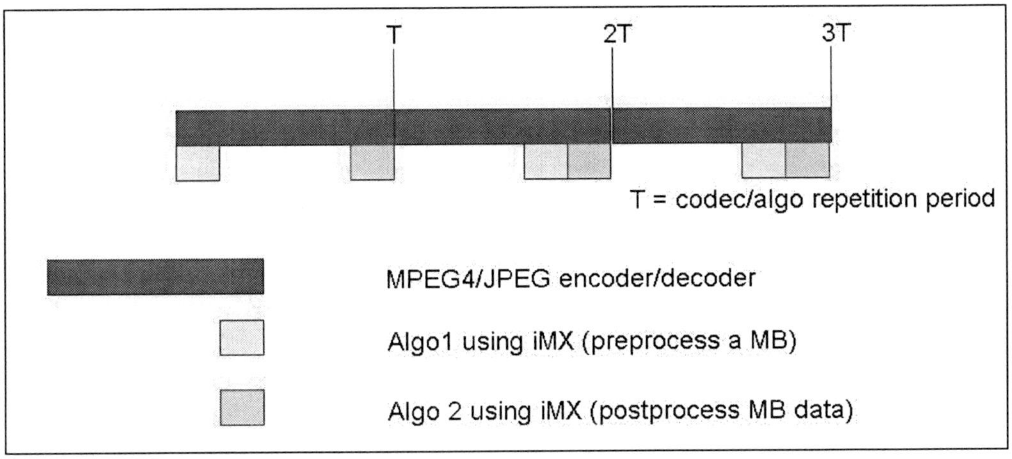

Figure 10.11: Concurrent execution of iMX programs with MPEG4/JPEG codecs.

DMA for IO

If additional DMA transfers are required due to the algorithm to fetch input or output, the DMA transfer will have to be chained/linked to the existing transfers in codec. This is needed to avoid control flow change within codec processing. The codec control flow is managed by COPC and SEQ. This would require codec source files (at least few of them). Codec will have to be revalidated.

Availability of iMX cycles

iMX free period may be used for iMX algorithms. (400 cycles of iMX is used per macroblock encoding. In other words, iMX is free for 70 to 84% of the codec execution time depending on encoder settings).

iMX program execution time

Currently, iMX is not the hardware block that determines codec execution time. codec execution time is determined by the worst-case block (in term of execution time), which are DMA and b-iMX. Thus, if the iMX execution time exceeds the execution time of the block that is the bottleneck for performance, codec performance will degrade and cause timing problems (since codecs are not tested for this case). Codec will need revalidation. Performance will have to be accounted by the application.

10.11 CONCLUSION

ARM9EJ is available for 40%–50% of total execution time to perform additional services. The MPEG4 and JPEG codecs use minimal ARM cycles as seen from datasheets. The ARM load per codec is less than 20 MHz. The rest of the codec processing is performed by MJCP. Scheduling of the additional services concurrently with MJCP (performing encode/decode) yields optimal utilization of ARM9E and MJCP. Additionally, ARM9EJ can be utilized when MJCP is finished encode/decode operation.

iMX can be used for pre/post processing algorithms operating on macroblock data of the frame that is being encoded or decoded respectively. In this case, care needs to be taken to ensure iMX program does not cause performance bottleneck for codecs (since codecs are not tested for this timing scenario). In addition, this concurrent operation requires the algorithms to be fitted within the available iMX program/data memory along with the iMX program of codecs.

If spare time is available after codec execution, iMX programs can be run sequentially with codecs. In this mode of operation, there is no limitation of program/data memory for algorithms as the entire hardware is available for the iMX program of algorithms. Also, other hardware modules like SEQUENCER, and EDMA can be used by the algorithm without many restrictions.

CHAPTER 11

Adding your secret sauce to the Signal Processing Layer (SPL)

11.1 INTRODUCTION

So far we have discussed the software platform that Texas Instruments provides and how you may develop your product based on it. However, some of you may have your own intellectual property that brings a unique differentiation to your product. While you may add this differentiation in the Application Layer (APL), depending on your algorithm, it may not run fast enough if the APL runs on the ARM processor alone. In such cases, you would want to leverage the power of the DSP on the SoC and migrate your algorithm to the DSP. *This chapter shows how you may componentize your secret sauce, port it to the DSP and integrate it with the rest of the software platform.*

The process of doing this is quite involved and we may not here be able to provide full justice to it. In the remainder of this chapter, I will try to touch on the key aspects. You may wish to get a more hands on experience by going through the teaching example posted online:

`http://www.ti.com/davinciwiki-portingtodsp`

11.2 FROM ANY C MODEL TO GOLDEN C MODEL ON PC

In most cases the starting point for your algorithm might be an ITU standards code or you may have developed your own code. This will most likely be in floating point, developed and tested on the PC. As a developer you will probably assume that you have access to unlimited amount of memory and processing power. Your goal would have been to first develop an algorithm that meets the needs of your application. The challenge now is to migrate it from the PC world to the embedded world where space (memory) and time (MHz) are limited due to the target cost goals of the end product. Your C code base must be modified to adhere to certain rules so that it can run in real time and easily be integrated with the rest of the software platform. Golden C is the resulting code base that follows these rules.

Step1: Create a test harness.

After confirming that the algorithm meets the needs of the target application, the first step now is to create a test harness with well defined input files, and corresponding output files that will be used later for verifying correctness of porting. This is an important step since you will be using the test files several times in the process of porting your code to the DSP.

Next, ideally, it would be better if the algorithm were converted from using floating point data types to fixed point data types. Depending on the type of SoC you choose from Texas Instruments, you may or may not have to do this. For example, there are devices with floating point support. However the majority of devices support only fixed point notation and if cost and power are your concerns then these fixed point devices become more attractive. While you may think that your algorithm must have floating point support, in general, it is possible to work around that requirement and develop an algorithm using fixed point arithmetic. There is a wealth of information and documentation available on how to convert your floating point to fixed point based processing and describing the steps involved in this process is outside the scope of this book.

Step 2: Convert your C to Golden C model.

In order to make your code componentized and meet real time and embedded processing requirements, it is better that it meet some basic requirements or rules enumerated below. While there are far many rules to follow, here are some key ones:

Rule1: Organize your code base. Organize your code based on functionality and place them in appropriate folders. While this might be obvious, source code organization, test files, and documentation allows multiple teams to work together and share their developments at a later stage in the development cycle. Test files should be isolated from the algorithm implementation.

Rule2: Your algorithm should not perform any file input and output operations. All data in and out should be via buffers using pointers for efficiency. Your top level application code may do file I/O operations but your core algorithm should never do them.

Rule3: Remove any mallocs and callocs. This is probably the hardest part of the conversion process since algorithm developers tend to use mallocs. However, for embedded processing, we want the framework such as the codec engine to be in charge of managing resources.

Rule4: Classify the type of memories used into persistent, scratch and constants. Persistent memory is memory that needs to be maintained from one frame to the next while scratch memory is scratch and need not be saved or maintained from one frame to the next. Constants are tables or coefficients that are needed by the algorithm. In large volume applications, constants can be moved to ROM thereby reducing the cost of the system.

Rule5: Avoid use of static and global variables.

Rule6. Data types should all be isolated into one header file with clear explanations. This will become useful later when you wish to leverage optimized code or kernels provided by TI and or third parties. Well defined data types that match the word length of the device and the library routine are important.

Rule7: Your code should not contain any endian specific instructions.

Rule8: Stack should be used only for local variables and parameter passing. Large local arrays and structures should not be allocated in the stack.

Rule9: Your code should be xDAIS-compliant. Tools are available for testing for this compliance.

Rule10: Your code should be xDM-compliant.

Step 3: Build, Run, and Test your Golden C code on PC

Now that you have made your code embedded processing friendly, build and test it using the test harness defined in step one to ensure that you did not introduce any new bugs! Benchmark your code on the PC to evaluate the performance. For example, you may wish to know the frames per second (fps) that you observe by running the code on the PC.

Step4: Build, Run and Test your Golden C code on the DSP using CCS.

Now you are ready to use Code Composer Studio (CCStudio), a software development tool and environment provided by Texas Instruments. Compile your code base for the DSP and use CCStudio to load it and run it on the DSP. You will reuse your test harness again to ensure that your code runs correctly on the DSP.

Step5: Basic DSP optimization using compiler options.

By turning on certain compiler options, you may be able to quickly see significant boosts in the performance of your code. Please see the wiki page shown in section 11.1 for details.

Step6: Make the code xDAIS and xDM compliant.

This is a necessary step in order to make your code integrate with the rest of the software from Texas Instruments and third parties. You may have already done this in Step2. While ideally Step2 is the correct stage at which you should be making your code xDAIS and xDM compliant, however, you may delay it to Step6. There are several tools that enable you to test your embedded code for compliance, which are not available to you on the PC.

Step 7. Create a server for Codec Engine.

Step 8. Test the server using DVTB as a reference example.

You can now use the Digital Video Test Bench (DVTB) code as a reference application for calling your xDM compliant algorithm from the ARM and measure the performance. You should observe a significant boost in the performance when your code runs on the DSP.

These 8 steps have provided you with a process for adding your unique differentiating features easily to the standard TI software platform.

Using the TI software platform, allows you to focus your creative energy more on your differentiation and less on the mundane tasks of writing basic software. You need not know all the complexity and details of the underlying hardware architecture in order to build your compelling product. The beauty of this is that you can take advantage of years of software engineering effort provided with TI's silicon and build on top of it.

CHAPTER 12

Further Reading

We hope that this book has given you an insight into the software platform and how it is organized and how you may develop applications based on it. In addition to this reading, there are several other resources and books that will accelerate your software development. Some of them are:

1. OMAP and DaVinci Software for Dummies, by Steve Blonstein and Alan Campbell. TI part #: SPRW184
2. DVTB documentation available online at `www.ti.com/davinciwiki_dvtb`
3. Porting GPP code to DSP and Codec Engine at `www.ti.com/davinciwiki-portingtodsp`
4. 3 different URLs wiki.davincidsp.com, wiki.omap.com and tiexpressdsp.com. The same content appears regardless of which URL you use. This was done in order to serve the needs of the Davinci^{TM}, OMAP^{TM}, and eXpressDSP^{TM} platforms without fracturing or duplicating content.
5. **Related Wiki and Project Sites**

    The Real-Time Software Components (RTSC) project wiki at eclipse.org

    TI Open-source projects

    Target Content Downloads (DSP/BIOS, Codec Engine, XDAIS, RTSC etc)

    Code Generation Tools Downloads

    Applications PowerToys Downloads

    TI DSP Village Knowledgebase
6. Leveraging DaVinci Technology for creating IPNetwork Cameras, TI Developers Conference, Dallas, Texas, March 14, 2007.
7. Accelerating Innovation with Off-the-Shelf Software and Published APIs, ARM Developers Conference '06, Santa Clara, CA, Oct 3, 2006. Invited presentation.
8. Accelerating Innovation with the DaVinci software code and Programming Model, TI Developers Conference, Dallas, Texas, Feb 28, 2006.

About the Author

BASAVARAJ I. PAWATE

Basavaraj I. Pawate (Raj) , Distinguished Member Technical Staff, has held several leadership positions for TI worldwide in North America, Japan, and India. These cover a wide spectrum of responsibilities ranging from individual research to initiating R&D programs, from establishing product development groups to outsourcing and creating reference designs, from winning designs and helping customers ramp to production to being CTO of emerging markets.

After completing his M.A.Sc. in signal processing at the University of Ottawa, Ottawa, Canada, Raj joined TI Corporate R&D in 1985 and worked on speech processing, in particular speech recognition for almost 10 years. He then moved to Japan where he established the Multimedia Signal Processing group from the ground up. When TI identified VoIP as an EEE, Raj went to Bangalore, India to establish a large effort in product R&D. Here he worked with Telogy, a company that Texas Instruments acquired, to deliver Janus, a multicore DSP device with VoIP software.

Raj is credited with several early innovations: a few examples include the world's first Internet Audio Player, a precursor to MP3 players, world wide standard for DSPs in standardized modules (Basava Technology), reuse methodologies for codecs and presilicon validation (CDR & Links & Chains), and one software platform for diverse hardware platforms.

Raj has fifteen issued patents in DSP algorithms, memory, and systems. Several of these patents have been deployed in products. Raj has published more than thirty technical papers.

Raj and his wife Parvathi have three daughters and live in Houston. Raj enjoys talking, walking, and, recently, reading philosophy.

Breinigsville, PA USA
09 June 2010
239421BV00002B/3/P

9 781598 299786